# Study Guide

# for

# Seeds'

# Horizons: Exploring the Universe

# Eighth Edition

## Mark A. Nook

*St. Cloud State University*

**THOMSON**

**BROOKS/COLE**

Australia • Canada • Mexico • Singapore • Spain • United Kingdom • United States

Printed in the United States of America
1   2   3   4   5   6   7   07   06   05   04   03

Printer: Victor Graphics, Inc.

ISBN: 0-534-39283-0

For more information about our products, contact us at:
**Thomson Learning Academic Resource Center**
**1-800-423-0563**

**For permission to use material from this text,**
**contact us by:**
**Phone:** 1-800-730-2214
**Fax:** 1-800-731-2215
**Web:** http://www.thomsonrights.com

**Brooks/Cole—Thomson Learning**
**10 Davis Drive**
**Belmont, CA 94002-3098**
**USA**

**Asia**
Thomson Learning
5 Shenton Way #01-01
UIC Building
Singapore 068808

**Australia/New Zealand**
Thomson Learning
102 Dodds Street
Southbank, Victoria 3006
Australia

**Canada**
Nelson
1120 Birchmount Road
Toronto, Ontario M1K 5G4
Canada

**Europe/Middle East/South Africa**
Thomson Learning
High Holborn House
50/51 Bedford Row
London WC1R 4LR
United Kingdom

**Latin America**
Thomson Learning
Seneca, 53
Colonia Polanco
11560 Mexico D.F.
Mexico

**Spain/Portugal**
Paraninfo
Calle/Magallanes, 25
28015 Madrid, Spain

# TABLE OF CONTENTS

# USING THE STUDY GUIDE

This study guide is designed to compliment *Horizons: Exploring the Cosmos* Eighth edition by Michael Seeds and to make your study of astronomy more efficient. This guide cannot take the place of reading the textbook or attending lecture and/or lab. It does complement each of these activities, and in particular it will focus your study of the textbook, provide you with worked examples of concept questions and problems involving mathematics, and provide a practice exam to help you gauge the extent to which you understand the material.

Most people taking this class are not science majors, and many would avoid science courses all together if they could. Science is different from many other disciplines and as a student you may need to spend more time studying astronomy than you expected. There is a proven mechanism for studying astronomy that follows the following basic outline.

     I. Read the Chapter
    II. Write out key terms (usually appearing in bold)
   III. Answer review questions on each section
   IV. Attend lecture, keep good notes, and stay engaged in the discussion
    V. Work homework problems
   VI. Take a practice Test
  VII. Review errors on homework and practice test
 VIII. Skim text and lecture notes before exam
   IX. Get a good night's sleep before each exam

Your first step in learning the material of the chapter is to read the chapter in the textbook. As you read, note the words printed in **bold**. These are terms essential to understanding concepts in the chapter. Make a list of these terms and make sure you understand what they mean. As you read the chapter, read the *REVIEW: Critical Inquiry* at the end of each section and answer the follow-up question. These *Reviews* serve as examples of how to answer essay questions and provide insight into concepts in the chapter.

Once you finish reading the chapter and listing the key terms, work through the questions in the **UNDERSTANDING THE CONCEPTS** section of the study guide. These questions are specific to the subsections in the textbook. They should help you develop an understanding of the important ideas and concepts presented in each subsection as well as each section. It is best to write out your answers to each of these questions. Writing down the answer takes time for you to think through each of the words as they are written and for most people greatly improves retention of the ideas. Answers to these questions are not provided, but answers are found in the subsection of the textbook that the questions are listed under.

The **KEY CONCEPTS** provide a brief summary of the most fundamental ideas presented in the chapter and often times point out how these concepts relate to other chapters in the text.

A few questions related to the concepts presented in the chapter are posed and then answered in the **QUESTIONS ON CONCEPTS** section. If you understand the concepts of the chapter, you should be able to answer these questions on your own. Solutions are provided so you can check your understanding. These questions and solutions provide further examples of concept related questions, similar to those in the *REVIEW: Critical Inquiry* boxes at the end of each section in the textbook.

One big concern of many students is understanding the mathematics. There is a great deal of variation between instructors as to the level of mathematics that will be required by students on homework, labs, and exams. You should ask you instructor the level of mathematics she/he requires in the course.

The **WORKED EXAMPLES OF PROBLEMS REQUIRING MATHEMATICS** will present at least one problem for each type of mathematics problem associated with the chapter. These examples present step-by-step procedures for solving problems. Reading through these examples will help you understand how to setup and complete these mathematics related problems.

At this point you are ready to attend the lecture. Most students don't get this order right. They attend lecture and then read the book. If you read the textbook and work through the **UNDERSTANDING THE CONCEPTS** questions before attending lecture, you will have some idea about what is being discussed. You will be more engaged in the topic and better able to understand the concepts the instructor presents. You will also be in a better place to answer questions. Most importantly, you will be prepared to ask questions on the material that does not make sense to you.

The lecture is crucial. Take good notes that indicate key concepts, not every little detail. You need to be able to follow what the instructor is saying and get the key points down. If you try to write down everything, you won't be following the logic of her/his arguments, and the lecture will be a blur. Jot down key points and ask questions about things that don't make sense.

Following lecture, complete any homework assignments and work through the end of chapter questions and problems. If mathematics is a part of the course, work through at least a few of the **MORE MATH RELATED PROBLEMS** in the study guide. Answers are provided for each of these at the end of the chapter in the study guide, so they provide a good way to determine if you know what you are doing when it comes to the math problems. If you are having trouble with these, consult with the instructor, teaching assistant, or a classmate.

It is important that you take the **PRACTICE TEST** honestly. Find a quiet spot and give yourself at least 30 minutes to complete the practice test uninterrupted. Do not use your textbook or any of your notes. Any question that you don't know the answer to, leave blank and mark that question. When finished correct the practice test against the answers at the end of the chapter of the study guide. Now go back and review any of the questions that you missed or left blank. Determine where that material is covered in the book and review it.

The **PRACTICE TEST** questions are not the ones that will appear on the exam; however, they are based on the material in the text, and they are similar to the types and wording of questions that are likely to be on the exam.

The night before an exam, skim through the text and review your lecture notes. Read through the practice test(s) and pay close attention to the material that gave you problems on the practice test(s). Then get plenty of sleep.

## CLOSING THOUGHTS

Many students treat their general education courses with contempt, and their grade shows it. Spending a little extra time reading the material and working through a few questions will help you understand the material much better, develop a greater appreciation for the subject, and above all *improve your grade!* I hope that you find this study guide helpful. If you have comments on the guide, I would like to hear from you. My intent is to help college students learn this material better and more efficiently, and I am always open to suggestions that will help students learn. So please send me you comments.

Thanks,

Mark

manook@stcloudstate.edu

# ACKNOWLEDGEMENTS

If this guide is at all useful, it is due primarily to the thousands of students that have taken astronomy classes with me during the past twenty-some years. Your questions and comments greatly shaped my understanding of what students have trouble comprehending and how I can make that material easier to grasp. Your constant call for more problems to work so that you could better understand the material gave me a great appreciation for your dedication to your education and your need to work with the concepts to gain a deeper comprehension of those concepts.

I have received a great deal of technical help in preparing this study guide. The people involved have helped me correct numerous errors, cumbersome usage, unreadable diagrams and graphs, and many other little things that greatly improve the readability and usefulness of this study guide. I am very appreciative of all the help that I received in this endeavor and thank them for their time and counsel.

Carol Benedict, the astronomy editor at Brooks/Cole, originally suggested this project and served as lead cheerleader through its production. She has been very supportive and was invaluable in enlisting the help of reviewers and overseeing the production of the guide. Thank you Carol.

Three people provided input on content and structure in this study guide. Paul Koblas at Western Arizona College and Robert A. Egler at North Carolina State University reviewed selected chapters and provided many helpful comments that made this a more usable guide for the students. John Banks offered much needed aid in properly formatting the mathematics that appears in the text, as well as offering suggestions for improving the author's editing skills. Thank you Paul, Robert, and John for your helpful advice.

Mehgan Nook reviewed the manuscript and helped to clean it up so that it's much easier to read. Thanks Mehgan for your careful reading and editing.

Paul Koblas, Robert Egler, John Banks, and Mehgan Nook have all greatly improved this study guide and any errors that still exist are solely my responsibility.

Preparing a study guide requires a great deal of time, which largely comes from time usually spent with one's family. My wife, Cheryl and our children have been very understanding of the time that it has taken to prepare this guide. Thank you Cheryl, Mehgan, Aaron, and Cory for your patience, understanding, and support.

# CHAPTER 1

# THE SCALE OF THE COSMOS

## UNDERSTANDING THE CONCEPTS

The table below provides a quick overview of the size of each image and the main idea presented.

| Image | Description | Ideas Presented | Diameter of View |
|-------|-------------|-----------------|------------------|
| 1-1 | Campus Scene | Scale | 52 feet |
| 1-2 | A Small City | Metric System & Distances | 1 mile (1.6 km) |
| 1-3 | Mountains and Rivers | Evolving Planet | 160 km |
| 1-4 | Earth | Earth's Structure | 12,756 km |
| 1-5 | The Lunar Orbit | Scientific Notation | $1.6 \times 10^6$ km |
| 1-6 | The Inner Solar System | Astronomical Unit (AU) | $1.6 \times 10^8$ km (1.07 AU) |
| 1-7 | The Solar System | Light Travel Time & Distance | 107 AU |
| 1-8 | Interstellar Space | Emptiness of Space | 11,000 AU |
| 1-9 | The Nearby Stars | Light-Year | $1.07 \times 10^6$ AU (17 ly) |
| 1-10 | Stars & Star Clusters | Star Clusters and Gas Clouds | 1,700 ly |
| 1-11 | The Milky Way Galaxy | Our Galaxy | 170,000 ly |
| 1-12 | The Local Group of Galaxies | Clusters of Galaxies | $1.7 \times 10^7$ ly |
| 1-13 | Filaments & Voids | Superclusters, Filaments, & Voids | $1.7 \times 10^9$ ly |

**1-1   CAMPUS SCENE**
> What is the typical size of objects in this image?

**1-2   A SMALL CITY**
> How many kilometers are in one mile?

**1-3   MOUNTAINS AND RIVERS**
> What evidence do we have that we live on a changing planet?  Think about answers beyond those mentioned in the text.
> Why is it important to use instruments other than just our eyes when we observe the natural world?

**1-4   EARTH**
> What causes the daily rising and setting of the sun and stars?
> What is the interior of Earth made of?

**1-5   THE LUNAR ORBIT**
> Write 82,500,000,000 in scientific notation.
> Write $4.29 \times 10^{-4}$ in standard notation.

**1-6   THE INNER SOLAR SYSTEM**
> What is a planet?
> What is a star?
> How many times larger than Earth is the sun?
> What is an astronomical unit?

**1-7   THE SOLAR SYSTEM**
> How much time does it take light from the sun to reach Mars?
> What is unique about Pluto's orbit?

**1-8   INTERSTELLAR SPACE**
> What objects are located within 10,000 AU of the sun?

**1-9   THE NEARBY STARS**
> How many astronomical units are there in one light-year?
> How far away is Proxima Centuri in astronomical units?

**1-10  STARS & STAR CLUSTERS**
What gives birth to new stars?
**1-11  THE MILKY WAY GALAXY**
What are the three major parts of our galaxy?
What holds our galaxy together and keeps the material form drifting apart and dissipating?
**1-12  THE LOCAL GROUP OF GALAXIES**
About how many galaxies are in the Local Group?
**1-13  FILAMENTS & VOIDS**
What are filaments?
What are voids?

## KEY CONCEPTS

The goal of the chapter is to illustrate the vastness of the universe through a series of images that gradually step further and further away from scenes that are part of our everyday experience. It is important that you understand the concept of scale. In the context used here, scale means a proportional increase or decrease. We are really looking at ratios, how many times larger or smaller is one thing than another. Do not think of the concept of scale being used only to distances. We will apply it to temperatures, the brightnesses of stars, the energy output of stars, the diameters and masses of stars and galaxies, etc.

This chapter provides an excellent brief overview of the structure of our universe: planets, solar systems, galaxies, clusters and superclusters of galaxies, and filament and void structure. The chapter also presents metric units and other units specific to astronomy (i.e. astronomical unit, light-year, parsec,). Appendix A in the text and Appendix 1 in this student guide present reviews of some mathematical material and some help in using calculators.

## WEBSITES OF INTEREST

http://www.powersof10.com/                    A closer look at the scale of things.

## WORKED EXAMPLES OF PROBLEMS REQUIRING MATHEMATICS

1.        What is $4.27 \times 10^8$ times $5.96 \times 10^{-2}$?

Solution:
Use your calculator for this. There are two things that you may not be use to that you have to get right and both involve the exponent.
First enter 4.27 then push the key labeled $\boxed{EE}$ or $\boxed{EXP}$ or $\boxed{EEX}$ and then push $\boxed{8}$.

The display should look like $\boxed{4.27 \quad 08}$ or $\boxed{4.27 \quad ^{08}}$ or $\boxed{4.27 \ E08}$ or $\boxed{427000000}$.

Push the multiplication key, $\boxed{\times}$.

Enter 5.96, push the $\boxed{EE}$, $\boxed{EXP}$, or $\boxed{EEX}$ key followed by the key labeled $\boxed{+/-}$ and then $\boxed{2}$.

The display should be $\boxed{5.96 \ -02}$ or $\boxed{5.96 \ ^{-02}}$ or $\boxed{5.96 \ E-02}$ or $\boxed{0.0596}$.

Finally push the equal key, $\boxed{=}$ and the answer should appear as

$\boxed{2.545 \quad 07}$ or $\boxed{2.545 \ ^{07}}$ or $\boxed{2.54 \ E07}$ or $\boxed{25449200}$.

So the answer is $2.54 \times 10^7$.

2. How many astronomical units are there in five light-years?

Solution:
From table A-6 in the appendix we find that
1 ly = $6.32 \times 10^4$ AU

The number of astronomical units in five light-years is found multiplying the number of light-years by the number of astronomical units in one light-year.

$$5 \text{ ly} = 5 \cdot (1\,\text{ly}) = 5 \cdot (6.32 \times 10^4 \text{ AU}) = 3.16 \times 10^5 \text{ AU}$$

So 5 ly = $3.16 \times 10^5$ AU = 316,000 AU.

3. How much time does it take light to travel from the sun to Saturn if it takes light 8.32 minutes to travel from the sun to Earth?

Solution:
The average distance from the sun to a planet is known as the semi-major axis of the planet's orbit. Table A-13 lists orbital information on the planets from which we find the following:
distance from the sun to Saturn = 9.54 AU and
distance the sun to Earth = 1.00 AU.
If it takes light 8 minutes to travel 1.00 AU, then it takes light 9.54 times 8.32 minutes to travel from the sun to Saturn.

$$t = 9.54 \text{ AU} \times \frac{8.32 \text{ minutes}}{1 \text{ AU}} = 79.4 \text{ minutes}$$

There are 60 minutes in one hour so

$$t = \frac{79.4 \text{ minutes}}{60 \frac{\text{minutes}}{\text{hours}}} = 1.32 \text{ hours}.$$

So it takes light 79.4 minutes or 1.32 hours to travel from the sun to Saturn.

4. The sun's diameter is $1.4 \times 10^9$ m and one astronomical unit is $1.5 \times 10^{11}$ m. How many suns laid side by side would be needed to stretch from the sun to Earth?

Solution:
We want to find how many suns it would take to cover a distance of $1.5 \times 10^{11}$ m. So the number of suns needed is equal to the distance to be covered divided by the size of one sun. Each sun covers $1.4 \times 10^9$ m.

$$number\ of\ suns = \frac{distance\ to\ be\ covered}{diameter\ of\ the\ sun} = \frac{1.5 \times 10^{11} \text{ m}}{1.4 \times 10^9 \text{ m}} = 107$$

So 107 suns laid side-by-side would be 1 astronomical unit long.

**MORE MATH RELATED PROBLEMS**
(Answers at the end of the chapter)

1.  Write 3,920,000 in scientific notation.

2.  What is the distance from the sun to Pluto in light-hours; the average distance for the sun to Pluto is 39.44 AU? Recall that the distance from the sun to Earth is 8.32 light minutes.

3.  In the diagram below, estimate the diameter of Uranus. Earth's radius is 6,378 km.

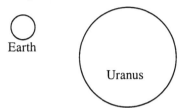

Earth

Uranus

4.  Work the following math problems using your calculator.

| Problem | | Answer |
|---|---|---|
| $(6.81\times10^{3})\cdot(3.29\times10^{8})$ | = | |
| $(2.83\times10^{-6})\cdot(7.15\times10^{4})$ | = | |
| $(4.37\times10^{-12})\cdot(8.35\times10^{-9})$ | = | |
| $(3.00\times10^{8})\cdot(4.91\times10^{-6})$ | = | |
| $(0.65\times10^{7})\cdot(0.0091)$ | = | |
| $(6.2\times10^{9})\div(3.1\times10^{4})$ | = | |
| $(2.43\times10^{12})\div(5.77\times10^{6})$ | = | |
| $(4.83\times10^{6})\div(1.52\times10^{14})$ | = | |
| $(5.32\times10^{5})\div(3.45\times10^{-8})$ | = | |
| $(8.33\times10^{-5})\div(3.00\times10^{8})$ | = | |
| $(7.12\times10^{-23})\div(9.22\times10^{-12})$ | = | |
| $(1.49\times10^{33})\div(8.39\times10^{-6})$ | = | |

5.  A parsec is equal to 206,265 AU. If light travels 1 AU in 8.32 minutes, how much time does it take light to travel 1 pc?
    Is this distance greater than a light-year?

**PRACTICE TEST**
**Multiple Choice Questions**

1.  The nearest star to the sun is
    a.  about 5 light-years away.
    b.  about 5 AU away.
    c.  about 50 light-years away.
    d.  about 50 AU away.
    e.  the North Star.

2. $4.28×10^8$ divided by $2.14×10^2$ is equal to
   a. $9.16×10^{10}$
   b. $2.00×10^6$
   c. $2.00×10^{10}$
   d. $9.16×10^6$
   e. none of the above

3. If Betelgeuse is at a distance of 520 light-years, then
   a. Betelgeuse is 520 million AU away.
   b. Betelgeuse must have formed 520 billion years ago.
   c. Betelgeuse is not part of the Milky Way galaxy.
   d. the light we see left Betelgeuse 520 years ago.
   e. none of the above.

4. Which of the following distances is the largest?
   a. The diameter of the sun
   b. The diameter of Earth.
   c. The diameter of Jupiter.
   d. 1 astronomical unit.
   e. The distance from Earth to the moon.

5. $3.65×10^{11}$ is the same as
   a. 3.65 thousand
   b. 365 thousand
   c. 3.65 billion
   d. 36.5 billion
   e. 365 billion

6. A unique feature of Pluto's orbit is that
   a. Pluto gets closer to the sun than any of the other planets
   b. Pluto's orbit carries Pluto halfway to the nearest star.
   c. from 1979 to 1999 Pluto was closer to the sun than Neptune.
   d. Pluto orbits the sun in the opposite direction compared to the other planets.
   e. Pluto's orbit is a perfect circle.

7. The Milky Way galaxy is part of a cluster of about 30 galaxies known as
   a. the Pleiades.
   b. a filament.
   c. a void
   d. an astronomical unit.
   e. the Local Group.

8. In the diagram at the right, what is the diameter of Mercury?
   a. about 400 km
   b. about 800 km
   c. about 1,600 km
   d. about 2,400 km
   e. about 3,600 km

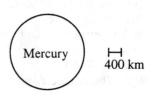

9. An astronomical unit is the average distance between
   a. stars.
   b. the sun and Earth.
   c. Earth and the moon.
   d. galaxies.
   e. clusters of galaxies.

10. The largest known structures in the universe are
    a. filaments.
    b. superclusters of galaxies.
    c. stars.
    d. black holes.
    e. galaxies

**Fill in the Blank**

11. A _____ is a small non-luminous body that shines by reflected light.

12. A _____ is equal to the distance that light travels in one year.

13. Write the number 78,346,000 in scientific notation.

**True False Questions**

14. The numbers $5.63 \times 10^{-3}$ and 0.000563 are equivalent.

15. A void is a region surrounded by long superclusters of galaxies known as filaments.

16. The average distance between stars is about 100 AU.

17. The Milky Way galaxy is a fairly small galaxy that lacks spiral arms.

18. A light-year is equal to the distance that light travels in one year.


**ADDITIONAL READING**

Morrison, Philip *Powers of Ten*. New York: W.H. Freeman, 1982.


**ANSWERS TO MORE MATH RELATED PROBLEMS**

1. $3.92 \times 10^6$
2. 328 minutes = 5.46 hours
3. 51,000 km

4.

| Problem | | Answer |
|---|---|---|
| $(6.81\times10^3)\cdot(3.29\times10^8)$ | = | $2.24\times10^{12}$ |
| $(2.83\times10^{-6})\cdot(7.15\times10^4)$ | = | $0.202 = 2.02\times10^{-1}$ |
| $(4.37\times10^{-12})\cdot(8.35\times10^{-9})$ | = | $3.65\times10^{-20}$ |
| $(3.00\times10^8)\cdot(4.91\times10^{-6})$ | = | $1470 = 1.47\times10^3$ |
| $(0.65\times10^7)\cdot(0.0091)$ | = | $59{,}150 = 5.92\times10^4$ |
| $(6.2\times10^9)\div(3.1\times10^4)$ | = | $2.0\times10^5$ |
| $(2.43\times10^{12})\div(5.77\times10^6)$ | = | $4.21\times10^5$ |
| $(4.83\times10^6)\div(1.52\times10^{14})$ | = | $3.18\times10^{-8}$ |
| $(5.32\times10^5)\div(3.45\times10^{-8})$ | = | $1.54\times10^{13}$ |
| $(8.33\times10^{-5})\div(3.00\times10^8)$ | = | $2.78\times10^{-13}$ |
| $(7.12\times10^{-23})\div(9.22\times10^{-12})$ | = | $7.72\times10^{-12}$ |
| $(1.49\times10^{33})\div(8.39\times10^{-6})$ | = | $1.78\times10^{38}$ |

5.  $1.72\times10^6$ minutes or 3.26 years.  A parsec is 3.26 times longer than a light-year.

**ANSWERS TO PRACTICE TEST**

1. a
2. b
3. d
4. d
5. e
6. c
7. e

8. d
9. b
10. a
11. planet
12. light-year
13. $7.8346\times10^7$

14. F
15. T
16. F
17. F
18. T

# C H A P T E R   2

# THE SKY

## UNDERSTANDING THE CONCEPTS

**2-1    THE STARS**
**Constellations**
What is the International Astronomical Union's definition of a constellation?
How many official constellations are there?
What is an asterism?
What is the central concept of this sub-section?
**The Names of the Stars**
What language was used to name the constellations?
In what language do most of the star names have their origin?
In addition to specific names, describe another system for naming stars.
What is the central concept of this sub-section?
**The Brightness of Stars**
In the magnitude scale, which star appears brighter, one of second magnitude or one of fourth magnitude?
What is meant by apparent visual magnitude?
What is the approximate magnitude of Polaris? (Use Figure 2-6)
What is the central concept of this sub-section?
**Window on Science 2-1.  Frameworks for Thinking About Nature: Scientific Models**
What is a scientific model?
Are they useful if they are not completely correct?
Can mathematical equations be used as scientific models?

What are the fundamental ideas presented in this section?

**2-2    THE SKY AND ITS MOTION**
**The Celestial Sphere and the Sky Around Us on pages 16 and 17**
How did the ancients view the sky?
Which direction does the sky appear to rotate?
What causes this apparent rotation of the sky?
Where is your zenith located?
How many seconds of arc are there in a minute of arc?
How many seconds of arc are there in a degree?
If you are standing on the equator, what do the star trails near the eastern horizon look like?
If you are in the southern hemisphere, what do star trails near the western horizon look like?
Where is Polaris located relative to your horizon if you are standing at Earth's equator?
What is the central concept of this sub-section?
**Precession**
What is precession?
How does precession affect the appearance of the night sky?
How long does it take Earth to complete one precession cycle?
**Window on Science 2-2.  Understanding Versus Naming: The True Goal of Science**
Why is the name of an object important?
Does knowing the name of an object, process or phenomena tell you anything about it?
What is the central concept of this Window on Science

What are the fundamental ideas presented in this section?

## KEY CONCEPTS

This chapter focuses on the appearance of the night sky. Many of the concepts presented were common knowledge before time became quantified on clocks and city lights blocked our nightly view of the sky. Most people today no longer have an understanding of the basic appearance or motions of the sky.

The two Window on Science (WOS) discussions in this chapter present concepts that are very important in all sciences. WOS 2-1 presents the concept of a scientific model. It is important to understand that scientific models do not have to be 100% accurate, 100% of the time to be useful. The discussion of the celestial sphere as a scientific model points out both the usefulness and limitations inherent in most scientific models.

The WOS 2-2 discusses the difference between naming something, or knowing the name of something, and understanding the process, object, or phenomena. An excellent example of this is a star. Most people can point to a star when asked what one is, but few can give a reasonably clear definition of one, describe what it is made of, how it produces energy, and/or how it changes with time. This illustrates that most people know that the name for one of those little points of light is "star", but their understanding of stars ends there. It is the goal of science to understand what objects, processes and phenomena are and how they operate. The names of objects, processes, and phenomena help us talk to each other about what we have learned and what we don't yet understand, but the central goal is a better understanding of how nature works.

One topic presented in this chapter that is confusing for many students is the magnitude of a star. It is important to understand this topic because magnitudes will be used in later chapters. The central concept is that magnitudes are related to brightness. However, the brighter the star, the smaller the magnitude associated with that brightness.

Finally, this chapter uses numerous diagrams and pictures to communicate critical information. It is important to study the figures and read the figure captions. Astronomy is a visual science and pictures greatly add to the understanding of a concept. Think of a picture as a miniature scientific model. It is a tremendous benefit to study and understand the diagrams, pictures, charts, and graphs in this section and throughout the text.


## WEBSITES OF INTEREST

| | |
|---|---|
| http://skyandtelescope.com/ | Lists events in the night sky |
| http://einstein.stcloudstate.edu/Dome/default.html | Online night sky with constellations |
| http://einstein.stcloudstate.edu/Dome/constellns/constlist.html | Greek Constellation Mythology |
| http://umbra.nascom.nasa.gov/eclipse/ | Eclipse information |
| http://aa.usno.navy.mil/data/docs/RS_OneDay.html | Sun and moon location data |

## QUESTIONS ON CONCEPTS

1.  An observer in the Northern Hemisphere takes a time exposure photograph of the night sky. If the illustration below depicts the photograph taken by the observer, which direction was the camera pointing: North, East, South, or West?

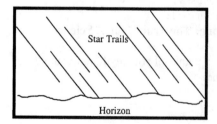

Solution:
The sky appears to rotate counter-clockwise around the north celestial pole for an observer in the northern hemisphere. If a northern hemisphere observer takes a time exposure facing north, they will record star trails that follow arcs appearing to go around the north celestial pole. If they take a time exposure facing south, they will record long arcs, arcing down toward the southern horizon. If they take the time exposure facing east, they will record long, nearly straight lines moving from the horizon toward the upper right. If they take a time exposure facing west, they will record long, nearly-straight lines moving from upper left to lower right.

This is a picture that could be recorded by a northern hemisphere observing facing the western horizon. What direction would a southern hemisphere observer be facing to record a similar image?
Answer: East. Can you explain why this is the case.

2.  If a star is on your zenith, where do you look to see it?

    Solution:
    The zenith is the point on the celestial sphere where a line from the center of Earth, extended through the observer, would intersect the celestial sphere. In short the zenith is the point on the celestial sphere that is directly above an observer.

    If a star is located on your zenith, you need to look straight up to see it.

3.  If you live at 28° south latitude, what is the angular distance between the southern horizon and the south celestial pole?

    Solution:
    The angular distance between the horizon and the celestial pole is always equal to the latitude of the observer. In the southern hemisphere, the south celestial pole will be visible above the southern horizon at an angle equal to the latitude.

    For an observer at 28° S, the south celestial pole will be 28° above the observer' southern horizon.


**WORKED EXAMPLES OF PROBLEMS REQUIRING MATHEMATICS**

1.  Using the data from Figure 2-6, how many time brighter is the sun than the full moon?

    Solution:
    From Figure 2-6, the sun's apparent visual magnitude is estimated to be –27, and the full moon's apparent visual magnitude is estimated to be –12. The difference in the two magnitudes is 15. From Table 2-1, we see that the ratio of their intensities is then 1,000,000 or 1 million.

    You can also use the formula

    $$\frac{I_{sun}}{I_{moon}} = (2.512)^{(m(\text{moon})-m(\text{sun}))} = (2.512)^{(-12--27)} = (2.512)^{(-12+27)} = (2.512)^{15} = 1,000,678$$

    The sun is approximately 1 million times brighter than the full moon.

2. Star ε Ind has a magnitude of 4.7, star α Ara has a magnitude of 2.9, star Σ2398 A has a magnitude of 8.9, and star Lacaille 8760 has a magnitude of 6.7.
    a. Which star is the brightest and which is the faintest?
    b. Which star is brighter Σ2398 A or α Ara, and how many times brighter?

Solution to part a:
The brightest stars have the smallest magnitudes. So α Ara is the brightest star followed by ε Ind, Lacaille 8760, and the faintest is Σ2398 A.

Solution to part b:
From part a we see that α Ara is brighter than Σ2398 A. The magnitude of α Ara is 2.9 and the magnitude of Σ2398 A is 8.9. The difference in the magnitudes is then 6.0. From Table 2-1 we see that a difference of 6 magnitudes corresponds to a ratio of 250 in intensity. Therefore α Ara is 250 times brighter than Σ2398 A.

This can also be determined using the magnitude formula for *By the Numbers 2-1* on page 13.

$$\frac{I_{\alpha\,\text{Ara}}}{I_{\Sigma\,2398\,\text{A}}} = (2.512)^{(m(\Sigma2398\text{A})-m(\alpha\,\text{Ara}))} = (2.512)^{(8.9-2.9)} = (2.512)^{6.0} = 251$$

$$I_{\alpha\,\text{Ara}} = 251 \times I_{\Sigma\,2398\,\text{A}}$$

## MORE MATH RELATED PROBLEMS
(Answers at the end of the chapter)

Use the following table to answer the next four questions

| Star | Apparent magnitude |
|---|---|
| Luyten 726-8A | 12.5 |
| τ Ceti | 3.5 |
| Rigel | 0.14 |
| Ross 128 | 11.1 |
| Fomalhaut | 1.15 |

1. Arrange the stars in order of brightest to faintest.

2. List the stars that are bright enough to be seen with the unaided eye.

3. What is the approximate ratio of the intensity of Rigel to Fomalhaut?

4. What is the approximate ratio of brightness of τ Ceti to Luyten 726-8A?

5. Archernar is 160,000 times brighter than Wolf 359. The apparent magnitude of Archernar is 0.5, what is the apparent magnitude of Wolf 359?

**PRACTICE TEST**
**Multiple Choice Questions**

1.  The star εEri has an apparent magnitude of 3.7 and the star ε Ind has an apparent magnitude of 4.7. ε Eri is about _____ time(s) _____ than ε Ind.
    a.  1.0      brighter
    b.  1.0      fainter
    c.  2.5      brighter
    d.  2.5      fainter
    e.  8.4      brighter

2.  An observer at Earth's North Pole would find
    a.  Polaris on the northern horizon.
    b.  Polaris 40° above the northern horizon.
    c.  that the ecliptic passing directly overhead.
    d.  that the celestial equator passing directly overhead.
    e.  that the ecliptic coincides with the horizon.

3.  An observer in the Southern Hemisphere takes a time exposure photograph of the night sky. If the illustration to the right depicts the photograph taken by the observer, which direction was the camera pointing?
    a.  straight north
    b.  straight east
    c.  straight south
    d.  straight west
    e.  straight up, directly overhead

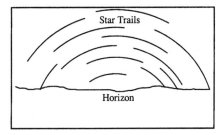

4.  61 Cyg has a magnitude of 5.2 and is about 6.25 times brighter than Sirius B. What is the approximate magnitude of Sirius B?
    a.  5.2
    b.  11.45
    c.  −1.05
    d.  3.5
    e.  7.5

5.  The precession of Earth
    a.  causes the seasons.
    b.  causes the location of the north celestial pole to move with respect to the stars.
    c.  causes the daily rising and setting of the stars, sun, and moon.
    d.  has a period of 365.25 days.
    e.  none of the above

6.  A northern hemisphere observer observes the circumpolar constellations
    a.  revolve clockwise around the north celestial pole.
    b.  revolve clockwise around the south celestial pole.
    c.  near the northern horizon.
    d.  near the southern horizon.
    e.  only during the winter.

7. You live at a latitude of 52° N. What is the angle between the northern horizon and the north celestial pole?
   a. 52°
   b. 48°
   c. 38°
   d. 23½°
   e. 90°

8. What information does a star's Greek letter designation convey?
   a. The apparent magnitude of the star.
   b. The time of year the star is visible
   c. The constellation in which the star is located.
   d. The brightness of the star relative to the other stars in that constellation.
   e. c and d

9. Which stars in the table to the right would be visible to the unaided eye of an observer on Earth?
   a. only α Cru and Lac 9352
   b. only Ross 154 and Lac 9352
   c. only α Cru, Deneb, and α CMa
   d. only α Cru and α CMa
   e. all of them

| Star Name | Apparent Visual Magnitude |
|---|---|
| Ross 154 | 10.6 |
| α Cru | 0.90 |
| Deneb | 1.26 |
| Lac 9352 | 7.4 |
| α CMa | −1.46 |

10. Which star in the table to the right appears the brightest to an observer on Earth?
    a. Ross 154
    b. α Cru
    c. Deneb
    d. Lac 9352
    e. α CMa

| Star Name | Apparent Visual Magnitude |
|---|---|
| Ross 154 | 10.6 |
| α Cru | 0.90 |
| Deneb | 1.26 |
| Lac 9352 | 7.4 |
| α CMa | −1.46 |

**Fill in the Blank Questions**

11. The magnitude scale can be used to measure the _____ of a star.

12. The _____ is the point on the celestial sphere directly opposite the zenith.

13. There are exactly _____ seconds of arc in one arc minute.

14. _____ of the rotation axis of Earth is caused by the force of gravity from the sun and moon on Earth's equatorial bulge.

15. If the north celestial pole appears on your horizon, what is your latitude?

**True-False Questions**

16. Star A has a magnitude of 5.0 and star B has a magnitude of 3.5, so star A is brighter than star B.

17. A scientific model is only useful if it is accurate for all situations.

18. The north celestial pole is located at the zenith for an observer on Earth's geographic north pole.

19. For an observer in the southern hemisphere, the angular distance of the south celestial pole above the southern horizon is equal to the observer's latitude..

20. There are 3600 seconds of arc in one degree.

## ADDITIONAL READING

Condos, Theony *Star Myths of the Greeks and Romans: A Sourcebook*, Grand Rapids, MI: Phanes Press, 1997.

MacRobert, Alan M. "Backyard Astronomy: Dissecting Light Pollution." *Sky and Telescope 92* (Nov. 1996), p. 44.

Monroe, Jean Guard, and Ray A. Williamson *They Dance in the Sky: Native American Sky Myths*. Boston: Houghton Mifflin, 1987.

Staal, Julius, D.W. *The New Patterns in the Sky: Myths and Legends of the Stars*. Blacksburg, VA: McDonald and Woodward Publishing, 1988.

Williamson, Ray A. *Living the Sky: The Cosmos of the American Indian*. Boston: Houghton Mifflin, 1984.

## ANSWERS TO MORE MATH RELATED PROBLEMS
1. Rigel, Fomalhaut, $\tau$ Ceti, Ross 128, Luyten 726-8A
2. Rigel, Fomalhaut, $\tau$ Ceti
3. 2.5
4. 4,000
5. 13.5

## ANSWERS TO PRACTICE TEST

| | | |
|---|---|---|
| 1. c | 9. c | 17. F |
| 2. a | 10. e | 18. T |
| 3. a | 11. intensity or brightness | 19. T |
| 4. e | 12. nadir | 20. T |
| 5. b | 13. 60 | |
| 6. c | 14. Precession | |
| 7. a | 15. 0° (you're at the equator) | |
| 8. e | 16. F | |

# CHAPTER 3
# CYCLES OF THE SKY

## UNDERSTANDING THE CONCEPTS

**3-1     THE CYCLE OF THE SUN**

**The Annual Motion of the Sun**

What is the ecliptic?

Through what angle does the sun appear to move in one day relative to the stars?

What is the central concept in this sub-section?

**The Seasons and The Cycle of the Seasons on pages 22 and 23**

What are the two causes of the seasonal variations in weather on Earth due to the tilt in Earth's rotation axis with respect to its orbital axis?

What is the autumnal equinox?

What is the winter solstice?

At what time of year for a northern hemisphere observer is Earth at its closest to the sun?

What is the zodiac?

What is the central concept in this sub-section?

**Window on Science 3-1. Astrology and Pseudoscience: Misusing the Rules of Science**

What is a pseudoscience?

Why is astrology considered a pseudoscience?

Read Appendix 2 of the Student Guide, for a more detailed discussion of pseudoscience and astrology.

**The Motion of the Planets**

Do planets produce their own visible light?

Which planets are visible without the use of binoculars or a telescope?

Viewed from above Earth's north celestial pole, which direction do the planets appear to orbit the sun?

What is the central concept in this sub-section?

What are the fundamental ideas presented in this section?

**3-2     THE CYCLES OF THE MOON**

**The Motion of the Moon**

Which direction does the moon move relative to the stars?

What is the central concept in this sub-section?

**Window on Science 3-2. Gravity: The Universal Force**

What do we mean when we say that gravity is universal?

What three factors determine how strong the gravitational attraction is between two objects?

**The Cycle of Phases and The Phases of the Moon on pages 28 and 29**

What causes the changing shape of the moon as it passes through its phases?

Why is there a difference between the period of the lunar phases and the orbital period of the moon?

At what time of day is it easiest to observe the first quarter moon?

What do the terms waxing and waning mean?

What is the central concept in this sub-section?

**Tides**
What causes the tides?
Why are there tides on opposite sides of Earth at the same time?
Which are stronger, tides caused by the moon or sun? Why?
When do spring and neap tides occur?
Which are larger, spring or neap tides?
What affect have tides had on the moon's rotation period?
What affect have tides had on the moon's orbital distance?
What is the central concept in this sub-section?
**Lunar Eclipses**
When does a lunar eclipse occur?
What is an umbra?
Why does the moon appear coppery red during a total lunar eclipse?
What is the central concept in this sub-section?
**Solar Eclipses**
During which phase of the moon does a solar eclipse occur?
What is an annular eclipse?
Why are some solar eclipses annular, while others are total?
What parts of the sun are visible during a total solar eclipse?
What is the central concept in this sub-section?
**Predicting Eclipses**
What is an eclipse season?
What are the nodes of the moon's orbit?
What is the Saros cycle?
What is the central concept in this sub-section?

What are the fundamental ideas presented in this section?

3-3     ASTRONOMICAL INFLUENCES ON EARTH'S CLIMATE
**The Hypothesis**
What is the Milankovitch hypothesis?
What three factors are mentioned that might influence Earth's long-term climatic variations?
What is the central concept in this sub-section?
**The Evidence**
What evidence suggests that the Milankovitch hypothesis might be correct?
What evidence suggests that the Milankovitch hypothesis might be incorrect?
What is the central concept in this sub-section?
**Window on Science 3-2. The Foundation of Science: Evidence**
Why is evidence so important in science?

What are the fundamental ideas presented in this section?


**KEY CONCEPTS**

This chapter focuses on the effects of the motions of Earth around the sun and the moon around Earth. The first section of the chapter describes the revolution of Earth around the sun and the tilt of Earth's rotation axis with respect to the orbital axis. From these descriptions the cause of the seasonal variations are explained. This tilt of Earth's rotation axis with respect to its orbital axis produces a variation in the angle at which sunlight hits the surface of Earth at a given location and produces the variation in the length of time that the sun is above the horizon at that location. During the summer the sunlight hits the northern hemisphere more directly and is above the horizon for a larger portion of the day than during the winter. This causes the northern hemisphere to warm significantly during the summer months.

The second section deals with the motions of the moon and the moon's changing appearance as a result of those motions. The first topic is the phases of the moon. It is the motion of the moon around Earth and the motion of Earth around the sun that produces the moon's changing phases and the difference between the synodic and sidereal months. Many students continue to believe that the phases of the moon are caused by the shadow of Earth on the moon – and this is not true. The changing phases are well described in *The Phases of the Moon* on pages 28 and 29.

The second major topic in the second section is lunar and solar eclipses and how to predict them. It is important to be able to mentally remove yourselves from the surface of Earth and look at the Earth-moon-sun system. This will help you understand the lunar phases and is critical to understanding how eclipses occur and why they don't occur every month. Being able to understand a drawing of a three-dimensional model is necessary. This is a skill that is needed to understand many of the diagrams of the moon's orbit and shadow. Additionally, three-dimensional figures will be used in every chapter that follows, so this is a good opportunity to develop your skill in reading three-dimensional diagrams.

Section 3-3 presents a good example of how science is done. The Milankovitch Hypothesis is presented and then analyzed in light of the evidence, both supporting and opposing. This example also demonstrates the inter relatedness of different disciplines.


**WEBSITES OF INTEREST**

        http://tycho.usno.navy.mil/srss.html                       Lunar phases at any date & time
        http://umbra.nascom.nasa.gov/                            Eclipse information


**QUESTIONS ON CONCEPTS**

1.    Why is it warmer in the summer than in the winter for someone living in the northern hemisphere?

    <u>Solution:</u>
    During the summer months, Earth's axis is tilted so that the sun is north of the celestial equator. This means that the sun will rise sooner and set later than when the sun is directly above or south of the celestial equator (i.e. during winter, spring, and fall). Since the sun will rise earlier and set later than in the winter, spring, and fall, the sun will be above the horizon for a longer period of time each day. This gives the sun more time to warm the air and Earth's surface during the summer months. Additionally the sun will be higher in the sky during the middle of the day. In the winter the sun is very low in the sky and solar radiation hits the northern hemisphere of Earth at a very oblique angle. During the summer the sun is higher in the sky and shines more directly on the northern hemisphere than it does during the winter, spring, and fall. Since the sun is shining more directly on the surface, it will more quickly warm that surface.

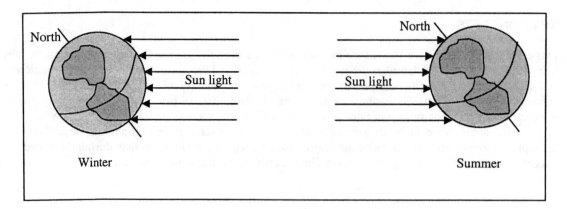

2.  Why can't Mercury and Venus ever be seen at midnight?

Solution:
Mercury and Venus have orbits that lie between the sun and Earth. This means that neither Mercury nor Venus can ever be than 90° from them sun, actually Mercury can never be more than about 24° and Venus about 45° from the sun, respectively. In order to be seen at midnight the object must be at least 90° away from the sun.

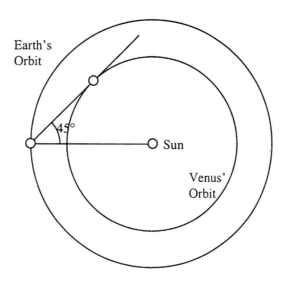

3.  Identify the phase of the moon on the first day of summer, if the moon is at the vernal equinox.

Solution:
In this questions we are asked to determine the phase of the moon on the first day of summer, when the moon is at the vernal equinox. This sounds a little confusing at first. When we say that the moon is at the vernal equinox, what we mean is that the moon is at the location relative to the stars, where the sun is on the vernal equinox. So if the moon is at the location of the vernal equinox, it is directly above Earth's equator at the point where the ecliptic crosses the equator. Each month the moon follows the ecliptic in the same manner the sun does in a year. In one month it moves from the location of the vernal equinox to the summer solstice, autumnal equinox, winter solstice and back to the autumnal equinox. If it is the first day of summer, the sun is at the summer solstice and the new moon will occur if the moon is at the summer solstice location as well. Progressing around the ecliptic, the first quarter moon will occur when the moon is at the autumnal equinox, followed by the full moon when it is at the winter solstice position, followed by the third quarter moon when the moon is at the vernal equinox. So on the first day of summer the third quarter moon will occur when the moon is at the vernal equinox location.

**WORKED EXAMPLES OF PROBLEMS REQUIRING MATHEMATICS**

1.  A marble has a diameter of 1.5 cm. It is located a long ways from you and you accurately measure its angular diameter to be 2 seconds of arc. How far away is the marble?

Solution:
We know:
the diameter of the marble = 1.5 cm, and
the angular diameter of the marble = 2.0 seconds of arc.

Since we know the linear diameter and angular diameter of the marble and want to find the distance, we should use the small angle formula in *By the Nu8mbers 3-1* on page 32.

$$\frac{angular\ diameter}{206{,}265} = \frac{linear\ diameter}{distance}$$

I want to find the distance to the marble so let me first invert both sides of the equation.

$$\frac{206{,}265}{angular\ diameter} = \frac{distance}{linear\ diameter}$$

I multiply both sides by the linear diameter and find that

$$distance = (linear\ diameter) \cdot \left( \frac{206{,}265}{angular\ diameter} \right) = (1.5\,\text{cm}) \cdot \left( \frac{206{,}265}{2''} \right) = 154{,}699\,\text{cm}$$

<u>A 1.5 cm marble with an angular diameter of 2 arc seconds is at a distance of 155,000 cm.</u>

I can convert this answer to a more convenient unit, namely kilometers. I know that there are 100 cm in a meter and 1000 m in a km, so there are 100×1000=100,000 cm in a km.

$$distance = 155{,}000\,\text{cm} = \frac{155{,}000\,\text{cm}}{100{,}000\,\frac{\text{cm}}{\text{km}}} = 1.55\,\text{km}$$

<u>A 1.5 cm marble with an angular diameter of 2 arc seconds is at a distance of 155,000 cm = 1.55 km, which is about 1 mile.</u>

2.  A quarter is about 1 inch in diameter. At what distance would a quarter have an angular diameter of equal that of the full moon?

<u>Solution:</u>
I know the diameter of the quarter and the angular diameter of the full moon
Diameter of the quarter = 1 inch
Angular diameter of the full moon = 0.5 degree = 0.5 degree × 3600 seconds of arc/degree = 1800 seconds of arc. (Note that we have used that there are 60 minutes of arc in a degree and 60 seconds of arc in a minute of arc. So there are 60 × 60 =3600 seconds of arc in one degree.)

We need to manipulate the small angle formula just like in the previous question so

$$distance = (linear\ diameter) \cdot \left( \frac{206{,}265''}{angular\ diameter} \right) = (1\,\text{inch}) \cdot \left( \frac{206{,}265''}{1{,}800''} \right) = 115\,\text{inches}$$

$$distance = 9\,\text{ft}\ \ 7\,\text{inches}$$

<u>So a quarter at a distance of 115 inches or 9 feet 7 inches will appear the same size as the full moon.</u>

## MORE MATH RELATED PROBLEMS
(Answers at the end of the chapter)

1.  What is the angular diameter of a dime at 100 meters? A dime has a diameter of 1.8 cm.

2.  What is the linear diameter of a ball whose angular diameter is 20 seconds of arc at a distance of 200 meters?

3.  Use the small angle formula to fill in the table below

|   | angular diameter | linear diameter | distance |
|---|---|---|---|
| **a** | 5" | 0.05 m | |
| **b** | | 71,500 km | $6.29 \times 10^8$ km |
| **c** | 1 minute of arc | | 5 miles |
| **d** | | $5.6 \times 10^8$ km | $4.9 \times 10^{15}$ km |
| **e** | 35" | | $1 \times 10^{21}$ m |
| **f** | 0.7" | $1.5 \times 10^{11}$ m | |

## PRACTICE TEST

### Multiple Choice Questions

1.  The _____ is 27.32 days long.
    a.  sidereal period of the moon
    b   synodic period of the moon
    c.  rotation period of Earth
    d.  revolution period of Earth
    e.  eclipse year

2.  As the moon orbits Earth
    a.  the same side of the moon always faces the sun.
    b.  the same side of the moon always faces Earth.
    c.  the moon will appear full when it is between Earth and the sun.
    d.  the moon moves about 1° per day relative to the stars.
    e.  the moon appears to move westward relative to the stars.

3.  The diagram below shows three approximate locations of the sun along the western horizon. Which number indicates the location of the sun at sunset on June 21 for an observer at a latitude of 48° S?

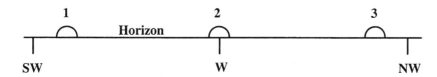

    a.  1
    b.  2
    c.  3
    d.  The sun will set in the east for an observer in the southern hemisphere.
    e.  The sun will not set on June 21 at this latitude.

4. During summer in the northern hemisphere
   I. sunlight shines more directly on the southern hemisphere than on the northern hemisphere.
   II. sunlight shines more directly on the northern hemisphere than on the southern hemisphere.
   III. the sun is above the horizon longer in the northern hemisphere than in the southern hemisphere.
   IV. the sun is above the horizon longer in the southern hemisphere than in the northern hemisphere.
   V. the sun passes through the zenith for all northern hemisphere observers.

   a. I, IV, & V
   b. II, III, & V
   c. I & III
   d. I & IV
   e. II & III

5. The sun is north of the celestial equator from the
   a. vernal equinox to the summer solstice.
   b. vernal equinox to the autumnal equinox.
   c. summer solstice to the winter solstice.
   d. autumnal equinox the vernal equinox.
   e. the sun is on the ecliptic and is never north of the celestial equator.

6. Which planets are never visible near the eastern horizon at sunset?
   a. Saturn and Mars
   b. Venus and Mars
   c. Jupiter and Mercury
   d. Mercury and Venus
   e. Jupiter and Saturn

7. The _____ rises at noon.
   a. sun on the summer solstice
   b. sun on the winter solstice
   c. new moon
   d. full moon
   e. first quarter moon

8. _____ when the moon is either new or full.
   a. Neap tides occur
   b. Spring tides occur
   c. Total solar eclipses occur
   d. Annular eclipses occur
   e. A coppery red moon will be visible

9. On the summer solstice the sun is
   a. 23½° north of the celestial equator.
   b. 23½° south of the celestial equator.
   c. on the celestial equator and moving north with respect to the equator.
   d. on the celestial equator and moving south with respect to the equator.
   e. closest to Earth.

10. If a total solar eclipse occurred in France in March of 1999. When was (will) this eclipse again (be) visible in France?
   a. September of 1999
   b. April of 2053
   c. April of 2017
   d. April of 1999
   e. Jan. of 2000

**Fill in the Blank Questions**

11. A(n) _____ is a set of beliefs that appears to be based on scientific ideas, but which fails to obey the most basic rules of science.

12. A(n) _____ eclipse occurs when the moon moves through the shadow of Earth.

13. The moon and visible planets are always within a few degrees of the _____.

14. For a northern Hemisphere observer the _____ _____ moon is visible in the south-eastern sky just before sunrise.

15. The _____ is the portion of Earth's or the moon's shadow where none of the light from the sun is visible.

**True-False Questions**

16. The first quarter moon rises at noon.

17. A total solar eclipse will occur somewhere on Earth whenever the moon is new.

18. For a person at a latitude of 42° N, the sun passes through the person's zenith at noon on the summer solstice.

19. Precession of Earth's axis causes the north celestial pole to move slowly relative to the stars.

20. The long-term changes in the temperature of Earth's ocean support the Milankovitch hypothesis.

## ADDITIONAL READING

Seeds, Michael A. "Astrology, UFOs and Pseudoscience." In *Astronomy Insights: Study Tips, Plus a Look at Astrology, UFOS*. Belmont, CA: Brooks/Cole Publising, 1997.
This article appears as Appendix 2 of this Study Guide.

Aveni, Anthony F. *Sky Watchers of Ancient Mexico*. Austin, TX: University of Texas Press, 1980, pp. 67-82.

Baliunas, Sallie and Willie Soon "The Sun-Climate Connection." *Sky and Telescope 92* (Dec. 1996), p. 38.

Carlson, Shawn "A Double Blind Test of Astrology." *Nature 318* (Dec. 5, 1985), p. 419.

Schneider, David "Trends in Climate Research: The Rising Seas." *Scientific American 278* (March 1997), p. 112.

## ANSWERS TO MORE MATH RELATED PROBLEMS

1. 37 seconds of arc
2. 0.0019 m = 1.9 mm
3.

|   | angular diameter | linear diameter | distance |
|---|---|---|---|
| **a** | 5" | 0.05 m | **2070 m** |
| **b** | **23.4"** | 71,500 km | $6.29 \times 10^8$ km |
| **c** | 1 minute of arc | **0.0015 miles = 7.7 feet** | 5 miles |
| **d** | **0.024"** | $5.6 \times 10^8$ km | $4.9 \times 10^{15}$ km |
| **e** | 35" | **$1.7 \times 10^{17}$ m** | $1.0 \times 10^{21}$ m |
| **f** | 0.7" | $1.5 \times 10^{11}$ m | **$4.4 \times 10^{16}$ m** |

## ANSWERS TO PRACTICE TEST

1. a
2. b
3. c
4. e
5. b
6. d
7. e
8. b
9. a
10. b
11. pseudoscience
12. lunar
13. ecliptic
14. waning crescent
15. umbra
16. T
17. F
18. F
19. T
20. T

# CHAPTER 4

# THE ORIGIN OF MODERN ASTRONOMY

## UNDERSTANDING THE CONCEPTS

### 4-1 PRE-COPERNICAN ASTRONOMY

**The Aristotelian Universe and The Ancient Universe on pages 44 and 45**

What were the two "first principles" that formed the basis of the geocentric model of the universe?

What is parallax?

Why couldn't ancient astronomers see parallax in the stars as Earth orbited the sun?

What is retrograde motion?

How is retrograde motion explained in the geocentric model of the solar system?

What is an epicycle and what is a deferent?

What are the fundamental ideas presented in this section?

### 4-2 COPERNICUS

**The Copernican Model**

What was Copernicus' great development to the model of the solar system?

What is meant by the term heliocentric universe?

**De Revolutionibus**

Why is *De Revolutionibus Orbium Coelestium* important?

What is a major advantage of the heliocentric model that Copernicus developed?

How is retrograde motion explained in the heliocentric model?

What is the major failure of the Copernicus' heliocentric model?

What aspect of the Aristotelian model did Copernicus maintain in his heliocentric system?

What are the central concepts of this sub-section?

**Window on Science 4-1. Creating New Ways to See Nature: Scientific Revolutions**

What is a scientific revolution?

What is a scientific paradigm?

Why are scientific paradigms important?

What are the fundamental ideas presented in this section?

### 4-3 TYCHO BRAHE

**The New Star**

What was the significance in the observation that the "new star" showed no parallax?

Why was Tycho able to make this discovery?

What is the central concept of this sub-section?

**Tycho Brahe's Legacy**

Why didn't Tycho accept the Copernicus' heliocentric model of the solar system?

What was Tycho's major accomplishment in astronomy?

**4-4    JOHANNES KEPLER**

**An Astronomer of Humble Origins**

As a young man, Kepler thought there could be only six planets, why?

**Joining Tycho**

What is the name of the astronomical tables that Kepler completed using Tycho's data?

The motions of which planet were studied by Kepler to first establish his three laws of planetary motion?

**Kepler's Three Laws of Planetary Motion**

What is Kepler's first law of planetary motion?

What is the eccentricity of an ellipse?

What is Kepler's second law of planetary motion?

What is Kepler's third law of planetary motion?

Kepler's third law can be written as $P^2 = a^3$ as long as $P$ is measured in _____ and $a$ is measured in _____.

What are the central concepts of this sub-section?

**The Rudolphine Tables**

What is the significance of the Rudolphine Tables?

**Window on Science 4-2.  Hypothesis, Theory, and Law: Levels of Confidence**

What is a hypothesis?

What is a model?

What is a theory?

What is a natural law?

What are the fundamental ideas presented in this section?


**4-5    GALILEO GALILEI**

**Telescopic Observations**

What were the three major discoveries Galileo reported in *Sidereus Nuncius*?

What is the significance of these discoveries in understanding the conflict between the geocentric and heliocentric models of the solar system?

What did Galileo discover about the surface of the sun, and why was it important in overthrowing the geocentric model.

What observations of Venus did Galileo make, and why were they important?

**Dialogo and Trial**

Why did Galileo face the Inquisition?

What are the fundamental ideas presented in this section?


**4-6    ISAAC NEWTON AND ORBITAL MOTION**

**Window on Science 4-3.  The Unstated Assumption of Science: Cause and Effect**

In what way does Newton's second law present a cause and affect relationship?

Why is cause and effect seen as the very foundation upon which science is based?

**Isaac Newton**

What is Newton's first law of motion?

What is Newton's second law of motion?

What is Newton's third law of motion?

Explain the difference between weight and mass.

What is meant when we say that the force due to gravity follows an inverse square relation?

Upon what three quantities does the force of gravity depend?

What do we mean when we say that gravitation is mutual?

What do we mean when we say that gravitation is universal?

What are the fundamental ideas presented in this sub-section?

**Orbital Motion and Orbiting Earth on pages 64 and 65**

What is a closed orbit?

What does the term center of mass mean?

Why is it correct to say that the moon is falling in its orbit?

What path will a satellite follow if its velocity is equal to the escape velocity?

Are astronauts traveling in the space station weightless? (Justify your answer.)

What are the central concepts presented in this sub-section?

**Window on Science 4-4. Testing a Theory by Prediction**

In what two ways are the predictions made by theories important?

**The Newtonian Universe**

In what ways were Newton's laws of motion and gravitation productive?

What are the fundamental ideas presented in this section?


## KEY CONCEPTS

The apparent major theme of this chapter is the history of astronomy, but within that is presented the method of science. This chapter demonstrates how science is done. Observations are made, hypothesis extended, theories and models constructed, and then they are tested. As new information and ideas come about, the models and theories are altered. Eventually, the hypothesis upon which they are based may be found to be invalid and new hypotheses, theories, and models must be constructed. Note that this process extends over generations of scientists.

The Greeks put forth a hypothesis and developed a model, the geocentric model, for the motion of planets and the structure of the universe. The geocentric model may not be an accurate description of nature, but it served to make the predictions the Greeks required and gave an excellent fit to the data they had. Copernicus suggested a different hypothesis and developed a model, the heliocentric model, that wasn't any better at fitting the observations available at that time. Tycho took very accurate data that allowed Kepler to make improvements in Copernicus' model. With these improvements, the heliocentric model more accurately described nature than the geocentric. Later, Newton showed that the laws developed by Kepler for motions of planets were controlled by physical principles that also controlled projectile motion on Earth.

Kepler's laws of planetary motion and Newton's laws of motion and gravity will be encountered in later chapters. These concepts are fundamental to understanding how we can measure the mass of a planet, star, or galaxy.

The four Window on Science presentations follow the process of science and help to make clear that the major topic of this chapter is the process of science, how it works, and why. Throughout this chapter and textbook we discuss theories, hypotheses, models, and laws. One key element of all of these is that they are constructed by us human beings to describe nature. This is especially important to keep in mind when thinking and talking about "laws of nature." Nature does follow certain rules and it is science's goal to try to understand those rules. We create theories and models that describe how we observe nature to work. But these "laws" are made by us and must remain open to constant revision. Nature does not follow our laws; we create these laws to describe what we see. As we gather more information and better understand the world around us, the theories, models, and laws will need to be modified. Einstein realized that the "Law of Gravity" did not adequately describe nature, especially when the gravitational field was strong. Einstein revised the law and made it applicable to a wider set of circumstances. The laws of physics are well established descriptions of how nature appears to work. They are testable, debatable, and they must be changeable in the light of further research that contradicts their predictions.

## WEBSITES OF INTEREST

http://www.wam.umd.edu/~tlaloc/archastro/      The Center for Archaeoastronomy
http://www.cv.nrao.edu/fits/www/yp_history.html      Great history of astronomy links
http://casswww.ucsd.edu/public/tutorial/History.html      History of Astronomy Tutorial
http://skyandtelescope.com/      Lists events in the night sky


## WORKED EXAMPLES OF PROBLEMS REQUIRING MATHEMATICS

1. If an object orbiting the sun has an average orbital distance of 15 AU, what is its orbital period?

<u>Solutions:</u>
This problem is an application of Kepler's third law.
I know that the average orbital distance = $a$ = 15 AU.

Kepler's third law is described on page 54 and states that $P^2 = a^3$ where $P$ is in years and $a$ is in astronomical units.

$$P^2_{years} = a^3_{AU} = 15^3 = 3375$$

If we take the square root of $P^2$ we obtain the period, $P$.

$$P_{years} = \sqrt{P^2_{years}} = \sqrt{3375} = 58 \text{ years}$$

<u>So an object with an orbital distance of 15 AU, has an orbital period of 58 years because $15^3 = 58^2$.</u>

2. If an object that orbits the sun has an orbital period of 340 years, what is its average distance from the sun?

<u>Solutions:</u>
This problem is an application of Kepler's third law.
I know that orbital period is 340 years.

Kepler's third law is described on page 54 states that $P^2 = a^3$ where $P$ is in years and $a$ is in astronomical units.

$$a^3_{AU} = P^2_{years} = 340^2 = 115,600$$

We can find average orbital distance, $a$, by taking the cube-root of $a^3$.

$$a_{AU} = \left(a^3_{AU}\right)^{1/3} = 115,600^{1/3} = 49 \text{ AU}$$

The last step requires that you take the cube-root of 115,600. Some calculators have a key labeled $\boxed{\sqrt[3]{\phantom{x}}}$ that will take the cube-root. Many others have a key labeled $\boxed{x^y}$ or $\boxed{x^{1/y}}$ both of which can be used to find the cube root.
To use the $\boxed{x^y}$ key, first enter the number 115,600, then push $\boxed{x^y}$ then enter 0.33333, and press the $\boxed{=}$ key. The display will give you $\boxed{\mathbf{48.71}}$, which is the cube-root of 115,600.
To use the $\boxed{x^{1/y}}$ key, first enter 115,600, then push the $\boxed{x^{1/y}}$ key, then enter $\boxed{3}$, and press the $\boxed{=}$ key. The display will give you $\boxed{\mathbf{48.71}}$, which is the cube-root of 115,600.

An object with an orbital period of 340 years, orbits at an average distance from the sun of 49 AU.

3.  A satellite orbits Mars 30,000 km from Mars' center. What is the satellites orbital velocity if it is in a circular orbit?

Solutions:
We know the orbital distance = $r$ = 30,000 km = (30,000 km)·(1,000 m/km) = $3.0 \times 10^7$ m.
The mass of Mars = $M$ = $6.42 \times 10^{23}$ kg.
The gravitational constant = $G$ = $6.67 \times 10^{-11}$ m³/s²kg.

$$V_C = \sqrt{\frac{GM}{r}} = \sqrt{\frac{\left(6.67 \times 10^{-11}\, m^3/s^2 kg\right) \cdot \left(6.42 \times 10^{23}\, kg\right)}{3.0 \times 10^7\, m}} = \sqrt{1.43 \times 10^6\, m^2/s^2} = 1,190 \tfrac{m}{s}$$

So the orbital speed of the object is 1,190 m/s or 1.19 km/s.

4.  Use the small angle formula to calculate the angular diameter of the Moon. The radius of the moon and the moon's distance from Earth can be found in table A-14 in the appendix.

Solutions:
The moon's linear diameter = 3476 km.
The average distance to the moon = 384,400 km.

We can use the small angle formula in *By the Numbers 3-1* on page 32 It is important when using the small angle formula that the linear diameter and the distance be in the same units, in this case km.

$$\frac{angular\ diameter}{206,265"} = \frac{linear\ diameter}{distance}$$

$$angular\ diameter = 206,265" \cdot \left(\frac{linear\ diameter}{distance}\right) = 206,265" \cdot \left(\frac{3476\ km}{384,400\ km}\right) = 206,265" \cdot (0.00904) = 1870"$$

$$angular\ diameter = 1870" \cdot \left(\frac{1\ minute\ of\ arc}{60\ seconds\ of\ arc}\right) = 31.1\ minutes\ of\ arc$$

$$angular\ diameter = 31.1\ minutes\ of\ arc \cdot \left(\frac{1\ degree}{60\ minutes\ of\ arc}\right) = 0.518°$$

The angular diameter of the moon is 1,870 seconds of arc, which is 31.1 minutes of arc, or about 0.518°. Recall that there are 60 seconds of arc in a minute of arc, and 60 minutes of arc in one degree.


## MORE MATH RELATED PROBLEMS
(Answers at the end of the chapter)

1.  A ball has a diameter of 3 inches (0.25 feet). At what distance would the ball have an angular diameter of 1 seconds of arc?

2.  At what distance would a star the size of our sun have an angular diameter of 0.1 second of arc? Express your answer in meters and astronomical units and take the sun's diameter to be $1.39 \times 10^9$ m.

3.  Betelgeuse has a linear diameter of $1.12 \times 10^{12}$ meters and a distance of 520 ly. What is the angular diameter of Betelgeuse as viewed from Earth? (1 ly = $9.5 \times 10^{15}$ m)

4.  A comet has an orbital period of 12.6 years, what is its average orbital distance from the sun?

5.  A satellite orbits the sun at a distance of 22 AU, what is its orbital period?

6.  What is the orbital velocity of Earth? The mass of the sun and Earth's orbital distance can be found in the appendix A-6 and A-5 respectively.

7.  The space shuttle orbits at an altitude of roughly 120 km, which is 120 km above Earth's surface. The radius of Earth is 6,380 km. What is the orbital velocity of the space shuttle if it follows a circular path?

**PRACTICE TEST**
**Multiple Choice Questions**

1.  Which of the following people accepted a geocentric model of the universe?
    a.  Kepler
    b.  Copernicus
    c.  Tycho
    d.  Galileo
    e.  Newton

2.  A _____ is a description of some natural phenomenon that can't be right or wrong. It is merely a convenient way to think about a natural phenomenon.
    a.  hypothesis
    b.  paradigm
    c.  natural law
    d.  model
    e.  theory

3.  The circular velocity of a satellite orbiting Earth depends on
    I.   the mass of Earth.
    II.  the mass of the satellite.
    III. the distance from the center of Earth to the satellite.
    IV.  the speed of light.

    a.  I & II
    b.  I & III
    c.  II & IV
    d.  I, II, & III
    e.  I, II, III, & IV

4.  Jupiter is on average about 5.2 AU from the sun. What is the approximate orbital period of Jupiter?
    a.  5.2 years
    b.  9 years
    c.  12 years
    d.  27 years
    e.  140 years

5. The period of Mars' orbit around the sun is approximately 1.9 years. What is the approximate distance from the sun to Mars?
   a.  6.9 AU
   b.  3.6 AU
   c.  2.6 AU
   d.  1.5 AU
   e.  1.3 AU

6. The _____ of an object is a measure of the gravitational force on an object.  On the other hand _____ is a measure of the amount of matter it contains.
   a.  weight        mass
   b.  mass          weight
   c.  energy        force
   d.  force         energy
   e.  momentum      energy

7. Thinking about Newton's laws of motion, the moon's elliptical path around Earth tells us that
   a.  a force is needed to keep the moon moving.
   b.  a force is needed to pull the moon away from straight-line motion.
   c.  the moon's orbital velocity does not depend on the mass of Earth.
   d.  the moon moves at a constant speed.
   e.  all of the above

8. Which of the following concepts was part of the Copernican theory for the solar system?
   a.  The sun was at the center of the solar system.
   b.  The planets followed elliptical paths around the sun.
   c.  Retrograde motion was described using equants and deferents.
   d.  all of the above
   e.  a and b

9. Planets generally move from west to east relative to the stars.  The outer planets occasionally stop and begin to appear to move from east to west.  The east to west motion is known as
   a.  parallax.
   b.  equant.
   c.  deferent.
   d.  parabolic motion.
   e.  retrograde motion.

10. Kepler's second law tells us that
   a.  a planet will travel faster when it is closer to the sun that when it is further from the sun.
   b.  a planet maintains a constant distance from the sun at all times.
   c.  the circular velocity of a planet depends on the sun's mass and the planet's distance from the sun.
   d.  objects orbiting the sun can not follow parabolic orbits.
   e.  parallax of stars can not be measured.

11. Classical Greek astronomers believed the motions of the heavens could be described by
   a.  elliptical motion.
   b.  the force of gravity.
   c.  retrograde motion.
   d.  uniform circular motion.
   e.  considering all planets to be within Earth's atmosphere.

**Fill in the Blank Questions**

12. _____ made some of the first telescopic observations of planets and found moon's orbiting Jupiter, mountains on the moon, and spots on the sun.

13. A(n) _____ is a system of rules and principles that can be applied to a wide variety of circumstances, but that is not universally accepted.

14. Given that Neptune is 30 AU from the sun, what is Neptune's orbital period?

**True-False Questions**

15. Tycho's observations of Mars were used by Kepler to develop Kepler's three laws of planetary motion.

16. Kepler's third law of motion relates a planets orbital speed to the planet's orbital period.

17. Parallax is the apparent motion of an object due to the motion of the observer.

18. Newton believed Earth could not move because he detected no parallax of the stars.

19. The Ptolemaic model of the universe is heliocentric.


**ADDITIONAL READING**

Gingerich, Owen *The Great Copernicus Chase and Other Adventures in Astronomical History.* Cambridge, MA: Sky Publishing Corp., 1992.

Hawking, Stephen W. and Roger Penrose "The Nature of Space and Time." *Scientific American 277* (July 1996), p. 60.

Musser, George "The Copernican Revolution Comes Around." *Mercury 25* (Sept./Oct. 1996), p. 14.


**ANSWERS TO MORE MATH RELATED PROBLEMS**
1. 51,600 ft = 9.8 miles
2. $2.87 \times 10^{15}$ m = 19,100AU
3. 0.047 seconds of arc
4. 5.4 AU
5. 103 years
6. 29,800 m/s = 29.8 km/s
7. 7830 m/s = 7.83 km/s


**ANSWERS TO PRACTICE TEST**

| | | | | | |
|---|---|---|---|---|---|
| 1. | c | 8. | a | 15. | T |
| 2. | d | 9. | e | 16. | F |
| 3. | b | 10. | a | 17. | T |
| 4. | c | 11. | d | 18. | F |
| 5. | d | 12. | Galileo | 19. | F |
| 6. | a | 13. | theory | | |
| 7. | b | 14. | 164 years | | |

# CHAPTER 5

# ASTRONOMICAL TOOLS

## UNDERSTANDING THE CONCEPTS

**5-1 RADIATION: INFORMATION FROM SPACE**
### Light as a Wave and a Particle
What is electromagnetic radiation?
What is a photon?
Which has a greater energy, a photon with a long wavelength or a photon with a short wavelength?
What is the central concept of this sub-section?
### The Electromagnetic Spectrum
What is a spectrum?
What is the wavelength in nanometers of red light and of blue light?
How do the wavelengths of x-rays compare to the wavelengths of visible light?
What are atmospheric widows?
What types of radiation pass through the atmospheric windows?
What is the central concept of this sub-section?

What are the fundamental ideas presented in this section?

**5-2 OPTICAL TELESCOPES**
### Two Kinds of Telescopes
What are the two principle components of a telescope?
What are the advantages of a reflecting telescope?
What is an achromatic lens and why is it used?
What is the central concept of this sub-section?
### The Powers of a Telescope
What is the light gathering power of a telescope?
What is resolving power of a telescope?
What causes seeing effects?
Why is light pollution a problem?
Why do astronomers place optical telescopes on selected high mountains?
What is the central concept of this sub-section?
### Window on Science 5-1. Resolving Power and the Accuracy of a Measurement
What effect often limits the resolution of astronomical images?
What is meant by the accuracy of a measurement?
### Buying a Telescope
When buying a telescope, why should we buy scopes with high quality optics?
Why should we buy a telescope with the largest possible diameter we can afford?
Why is the magnification of the telescope not a concern when buying a telescope?
What is the central concept of this sub-section?
### New-Generation Telescopes and Modern Astronomical Telescopes on pages 78 and 79
What were the three problems with traditional large telescope mirrors?
What two ways do astronomers now use when building large telescope mirrors to reduce the telescope's weight?
What is a segmented mirror?
What is a sidereal drive?
What is active optics?
What is the central concept of this sub-section?

### Interferometry
What is the advantage of interferometry?
What is an interferometer?
What is the central concept of this sub-section?

What are the fundamental ideas presented in this section?

**5-3**  SPECIAL INSTRUMENTS
### Imaging Systems
What is a CCD?
What are the advantages of a CCD compared to a photographic plate?
What is the central concept of this sub-section?
### The Spectrograph
What is a spectrograph?
What does a grating do?
How do astronomers use comparison spectra?
What is the central concept of this sub-section?

What are the fundamental ideas presented in this section?

**5-4**  RADIO TELESCOPES
### Operation of a Radio Telescope
What are the four major parts of a radio telescope?
What does the antenna on a radio telescope do?
What is the central concept of this sub-section?
### Limitations of the Radio Telescope
What are the three handicaps that radio astronomers must deal with?
In addition to size of the objective, what other factor influences the resolving power of a telescope?
Why is a radio interferometer useful?
What is the VLA?
Why are radio telescopes located in valleys and not on the tops of mountains?
What is the central concept of this sub-section?
### Advantages of Radio Telescopes
What are the three principle advantages of radio telescopes?
What is the central concept of this sub-section?

What are the fundamental ideas presented in this section?

**5-5**  SPACE ASTRONOMY
### Infrared Astronomy
What is the main absorber of infrared radiation in Earth's atmosphere?
What is SOFIA and SIRTF?
Why must infrared telescopes and detector be cooled?
What is the central concept of this sub-section?
### Ultraviolet Astronomy
What is the main absorber of ultraviolet radiation in Earth's atmosphere?
How have we been able to make observations at ultraviolet wavelengths if ultraviolet radiation doesn't pass through Earth's atmosphere?
What is the central concept of this sub-section?
### High Energy Photons
What sorts of astronomical objects produce x-rays?
What is Chandra?
What are gamma-rays?
What is the central concept of this sub-section?

**The Hubble Space Telescope**

What are two advantages to having a large telescope in orbit?

What is the central concept of this sub-section?

What are the fundamental ideas presented in this section?


## KEY CONCEPTS

The emphasis of this chapter is on the tools that astronomers use to study the universe. It is important to realize that telescopes by themselves have very limited use. The instruments such as spectrographs, CCD detectors, and photometers and the techniques such as interferometry make telescopes much more useful. Most of our understanding of the physical processes of the universe has come about because of the data collected with these instruments attached to telescopes.

The first section presents the electromagnetic spectrum. It is important to understand that all of the various types of radiation discussed in the remainder of the chapter are the same physical process and only their wavelengths/frequencies/energies are different. In astronomy we cannot do real experiments. We must simply observe the objects of the universe and all the information that we obtain about them comes via electromagnetic radiation. The material from this section will be valuable in understanding chapter 6 and how astronomers are able to study and understand phenomena in the universe.

The important point in the second section is that telescopes have different designs that allow us to build telescopes to perform certain tasks most efficiently. Each design has its advantages and disadvantages.

The third section describes the principal instruments astronomers use to record data from all celestial objects. The CCD camera has revolutionized astronomical imaging, because of its extreme sensitivity and ability to record faint and bright objects at the same time. The CCD also allows us to record images remotely, making the Hubble Space telescope and other orbiting telescopes possible. The spectrograph is the most important instrument in astronomy and over 80% of all allocated telescope time is for spectroscopy of one form or another. The spectrograph allows us to analyze the various wavelengths of light to determine nearly any aspect of a star's, planet's, or nebula's characteristics.

The last two sections describe the importance of using ALL of the light that objects emit. Radio and space observations have opened up many new ways to study and understand the solar system and cosmos. It has only been in the last 20 years that we have really begun to utilize the far infrared, ultraviolet, x-ray and gamma-ray radiation that astronomical objects emit. These observations have greatly changed our understanding of individual objects from great clouds of gas and dust, to supernovae, neutron stars, and the very structure of the universe.


## WEBSITES OF INTEREST

| | |
|---|---|
| http://www.stsci.edu/ | Hubble Space Telescope |
| http://www.noao.edu/ | National Optical Astronomy Observatory |
| http://www.nrao.edu/ | National Radio Astronomy Observatory |
| http://www.aao.gov.au/ | Anglo-Australian Observatory |
| http://www.ifa.hawaii.edu/ifa/ | Institute for Astronomy at the University of Hawaii |

**QUESTIONS ON CONCEPTS**

1. Why are optical telescopes built on the tops of mountains?

   Solution:
   Regardless of the type of telescope being built, it needs to be placed where it can collect the best data possible, in the most efficient manner. Optical telescopes are limited by the quality of the atmosphere and by light pollution. Light pollution is caused by city lights and other man made lights. Most cities are well below the tops of most 10,000 ft mountains, so mountaintops generally help avoid light pollution.

   The other major concern is atmospheric quality, or seeing. The atmosphere disturbs the light as it passes through it. The atmosphere above high mountains is thin and often times the airflow is smooth. This reduces atmospheric turbulence and improves the seeing. Mountaintops are more frequently above clouds as well. Therefore, high elevations provide darker skies, clearer skies, and skies less affected by atmospheric distortions.

2. Why are charge-coupled devices so useful in astronomy?

   Solution:
   In most cases in astronomy we are trying to measure the amount of light coming from very faint objects. We use large telescopes to gather as much of that light as possible, but we still have very little light collected from most objects. Consequently, we need detectors that are as sensitive as possible. CCDs have two big advantages for use in astronomy, they are very sensitive and they are able to record images of very bright and very faint things at the same time. Photographic plates are not very sensitive to light and typically have to be exposed for a very long time to record images of galaxies, nebulae and other faint astronomical sources. CCDs greatly reduce this exposure time and allow astronomers to collect higher quality data in a shorter amount of time. Therefore, on any given night astronomers can collect data on more sources with a CCD, than they can with photographic plates. Secondly, photographic plates work well for recording multiple objects in a single exposure only if the two objects are about the same brightness. Images on photographic plates of objects with greatly different brightnesses, either overexpose the brighter object, or underexposed the fainter object. In either case that information is unusable. CCDs have what is called a large dynamic range. This allows them to record data on very bright and very faint sources at the same time on a single image.

**WORKED EXAMPLES OF PROBLEMS REQUIRING MATHEMATICS**

1. What is the wavelength of radiation with a frequency of $1.5 \times 10^7$ Hz, and what type of radiation is this?

   Solution:
   A problem of this type appears as problem 2 on page 91. The exact equation used to solve the problem is not given, however, it is possible to figure it out. First, let's write down what we know.
   Frequency of radiation $= f = 1.5 \times 10^7$ Hz,
   Speed of light $= c = 3.0 \times 10^8$ m/sec.
   We want to find the wavelength $= \lambda$.

   We know that the frequency of a wave is the number of waves that pass a given point in one second. So if the waves all have a length of 10 ft, and were traveling at 2 ft/sec, how many would pass by a point each second? That seems easy, the frequency would be 10 ft divided by 2 ft/sec or 5 waves, that is 5 Hz. This implies that the frequency of a wave is equal to the wavelength divided by the speed, that is

   $$f = \frac{c}{\lambda}$$

In the problem we are presented, $f = 1.5 \times 10^7$ Hz and $c = 3.0 \times 10^8$ m/sec. Substituting in to the above equation we find

$$1.5 \times 10^7 = \frac{3.0 \times 10^8 \frac{m}{s}}{\lambda}.$$

If I multiply both sides by $\lambda$,

$$f \cdot \lambda = \lambda \cdot \frac{c}{\lambda} = c.$$

I now divide both sides by f, to can find an expression for $\lambda$,

$$\frac{f \cdot \lambda}{f} = \lambda = \frac{c}{f}. \qquad \text{So} \qquad \lambda = \frac{c}{f}.$$

Substituting in the given values, we find

$$\lambda = \frac{3.0 \times 10^8 \frac{m}{s}}{1.5 \times 10^7 \text{ Hz}} = 20 \text{ m}.$$

Radiation with a frequency of $1.5 \times 10^7$ Hz has a wavelength of 20 m which is a radio wave.

2. Compare the light gathering power of a 2-inch telescope to that of the 200-inch telescope.

Solution:
A 2-inch telescope has an objective with a diameter of 2 inches, while a 200-inch telescope has an objective with a diameter of 200 inches. The light gathering power of a telescope depends on the diameter of the objective squared, see *By the Numbers 5-1*, page 77.

Diameter of larger telescope = $D_A$ = 200 inches.
Diameter of smaller telescope = $D_B$ = 2 inches.
We want to find the ratio of the light gathering power of the larger telescope, $LGP_A$, to the light gathering power of the smaller telescope, $LGP_B$.

$$\frac{LGP_A}{LGP_B} = \left( \frac{D_A}{D_B} \right)^2 = \left( \frac{200 \text{ inches}}{2 \text{ inches}} \right)^2 = (100)^2 = 10,000$$

This tells us that the light gathering power of a 200-inch telescope is 10,000 times greater than the light gathering power of a 2-inch telescope. Put another way, a 200-inch telescope would collect as much light in one second as a 2-inch telescope would in 10,000 seconds (i.e. in 2.8 hours).

3. What diameter of a telescope is needed to provide a resolving power of 0.75 seconds of arc

   Solution:
   The resolving power is discussed in *By the Numbers 5-1* on page 77. The formula is

   $$\alpha = \frac{11.6}{D}$$ where $\alpha$ is the resolving power in seconds of arc and $D$ is the diameter in centimeters.

   We need to solve this for the diameter, $D$, so multiply both sides by $D$ and then divide by $\alpha$.

   $$D \cdot \alpha = D \cdot \frac{11.6}{D} = 11.6.$$ Now dividing by $\alpha$ we find

   $$D = \frac{11.6}{\alpha} = \frac{11.6}{0.75} = 15.5 \, cm.$$

   A telescope with a diameter of 15.5 cm will provide a resolving power of 0.75 seconds of arc. To achieve this resolving power, the telescope would have to be placed outside Earth's atmosphere.

4. A telescope has a mirror that is 0.2 meters in diameter and a 2-meter focal length. What focal length eyepiece would you need to obtain a magnification of 150× with this telescope?

   Solution:
   The magnifying power of a telescope and eyepiece is discussed in *By the Numbers 5-1* on page 77. The formula for magnification is given by

   $$M = \frac{F_{objective}}{F_{eyepiece}} = \frac{F_o}{F_e}.$$

   If we multiply both sides by $F_e$ and then divide both sides by $M$ we find

   $$F_e \cdot M = F_e \cdot \frac{F_o}{F_e} = F_o$$
   Then dividing by $M$

   $$\frac{F_e \cdot M}{M} = \frac{F_o}{M}.$$

   Canceling the $M$ on the left side of the equation we can solve for $F_e$.

   $$F_e = \frac{F_o}{M} = \frac{2.0 \, m}{150} = 0.0133 \, m = 1.33 \, cm$$

   An eyepiece with a focal length of 0.0133 m (1.33 cm) would provide a magnification of 150× with a 2.0 meter focal length telescope. The diameter of the telescope is not important in determining the magnification.

## MORE MATH RELATED PROBLEMS
(Answers at the end of the chapter)

1.  What is the resolving power of a telescope with an 8-inch diameter objective? (1 inch =2.54 cm)

2.  What diameter telescope is needed to be able to resolve two stars that are 0.03 seconds of arc apart?

3.  What is the resolving power of a typical set of binoculars with 35-mm objective lenses?

4.  What is the ratio of the light gathering power of a typical set of binoculars with 35 mm objectives lenses to that of the human eye with a pupil diameter of 7 mm?

5.  What is the ratio of the light gathering power of an 8-inch diameter telescope to that of the human eye with a 7-mm diameter pupil?

6.  An 8-inch diameter Schmidt-Cassegrain telescope has an effective focal length of 200 cm. What is the magnification if this telescope is used with a 25 mm eyepiece?

7.  The magnification of a typical pair of binoculars is 7×. If the focal length of the objectives is 6 inches, what is the focal length of the eyepieces?

8.  You are asked to design a telescope with a resolving power of at least 0.05 arc seconds.
    a.  What is the minimum diameter of the objective?
    b.  If the focal length of the objective is 5 times greater than the diameter of the objective, what is the focal length of the objective?
    c.  What size eyepiece would you need to use to obtain a magnification of 100×?


## PRACTICE TEST
## Multiple Choice Questions

1.  What is the wavelength of the shortest wavelength light that can be seen with the human eye?
    a.  400 nm
    b.  4000 nm
    c.  7000 nm
    d.  700 nm
    e.  $3 \times 10^8$ m

2.  _____ have wavelengths that are longer than visible light.
    I.   radio waves
    II.  Ultraviolet light
    III. Infrared radiation
    IV.  X-rays

    a.  I & III
    b.  II & IV
    c.  II & III
    d.  II, III, & IV
    e.  I, II, & IV

3.  Which of the following types of electromagnetic radiation has the greatest energy?
    a.  X-rays
    b.  Visible light
    c.  Ultraviolet
    d.  Radio waves
    e.  Infrared radiation

4. Why are radio telescopes located in valleys?
   a. The walls of the valley protect the telescopes from streetlights in nearby cities.
   b. The walls of the valley help collect the radio waves.
   c. The walls of the valley help shield the telescope from man-made radio signals.
   d. The atmosphere is more transparent to radio waves in valleys than it is on mountaintops.
   e. Radio telescopes aren't placed in valleys, they are placed on high mountaintops.

5. A telescope _____ will have a very large light gathering power.
   a. with a long focal length objective
   b. with a large diameter objective
   c. with a spectrograph attached to it
   d. with a short focal length eyepiece
   e. located on a mountaintop

6. The resolving power of an optical telescope with a diameter of 58 cm is
   a. 11.6 seconds of arc.
   b. 58 seconds of arc.
   c. 2.32 seconds of arc.
   d. 5 seconds of arc.
   e. 0.2 seconds of arc.

7. What is the ratio of the light gathering power of a 4-m telescope to that of a 2-m telescope?
   a. 2 to 1
   b. 1 to 2
   c. 4 to 1
   d. 16 to 1
   e. 1 to 16

8. _____ is absorbed by ozone high in Earth's atmosphere and requires that telescopes for observing at these wavelengths be placed in space.
   a. Infrared radiation
   b. Ultraviolet radiation
   c. Radio wave radiation
   d. Visible light
   e. none of the above

9. Why are the new large telescopes made using thin mirrors instead of the conventional thick mirrors?
   I. Thin mirrors cool faster at nightfall.
   II. Thin mirrors are cheaper because they require less glass.
   III. Thin mirrors require less time to grind them.
   IV. It is easier to change the mirror's shape to compensate for atmospheric seeing.
   a. I, II, & III.
   b. I, III, & IV.
   c. II & IV.
   d. II, III, & IV.
   e. I, II, III, & IV.

10. A(n) _____ is an astronomer's most powerful tool. It spreads light from an object out so that the intensity of light at various wavelengths can be studied.
    a. photometer
    b. charge coupled device
    c. interferometer
    d. spectrograph
    e. very long baseline array

**Fill in the Blank Questions**

11. A _____ telescope has an objective that is a mirror.

12. A(n) _____ has a few million light sensitive diodes in an array typically about a half-inch square.

13. _____ optics is a method for controlling the shape of a telescope mirror to help correct for the affects of atmospheric seeing.

14. A(n) _____ is used to move a telescope slowly so that it can follow a celestial object as Earth rotates.

15. The technique of connecting multiple telescopes together to combine the images from each telescope is known as _____.

**True-False Questions**

16. Photons of ultraviolet radiation have higher energies than photons of visible light.

17. A telescope with a long focal length will have a greater light gathering power than a telescope with a shorter one.

18. The prime focus allows larger instruments to be mounted on a telescope than does the Cassegrain focus.

19. The Very Large Array (VLA) is a radio interferometer.

20. Active optics are used to control the shape of a floppy or segmented mirror as the telescope is pointed in different directions.

**ADDITIONAL READING**

Hajian, Arsen R. and J. Thomas Armstrong "A Sharper View of the Stars." *Scientific American 284* (March 2001), p. 56.

Jayawardhana, Ray "Probing Cosmic Depths." *Astronomy 28* (Sept. 2000), p. 46.

Lazio, T. Joseph W., "Razor-Sharp Radio Astronomy." *Mercury 30* (May/June 2001), p. 32.

Perryman, Michael "Hipparcos:  The Stars in Three Dimensions." *Sky and Telescope 97* (June 1999), p. 40.

Sincell, Mark "Making Stars Stand Still." *Astronomy 28* (June 2000), p. 42.

Zimmerman, Robert "Seeing with X-ray Eyes" *Astronomy 29* (May 2001), p. 36.

**ANSWERS TO MORE MATH RELATED PROBLEMS**

1. 0.57 seconds of arc
2. 3.9 m
3. 3.3 seconds of arc
4. 25
5. 840
6. 80×
7. 6/7 inches = 0.86 inches
8. a. 232 cm = 2.32 m,
   b. 1160 cm =11.6 m
   c. 11.6 cm focal length eyepiece

**ANSWERS TO PRACTICE TEST**

1. a
2. a
3. a
4. c
5. b
6. e
7. c
8. b
9. e
10. d
11. Reflecting
12. charge coupled device
13. Adaptive
14. sidereal drive
15. interferometry
16. T
17. F
18. F
19. T
20. T

# CHAPTER 6

# STARLIGHT AND ATOMS

## UNDERSTANDING THE CONCEPTS

**6-1** ATOMS

**A Model Atom**

What are the major components of an atom?

What particles are found in the nucleus?

What are the relative characteristics (mass and charge) of protons, neutrons, electrons, and a nucleus?

What is the central concept in this sub-section?

**Different Kinds of Atoms**

What is an isotope?

What is an ion?

What is the difference between an ion and an isotope?

What is meant by ionization?

How is a molecule different from an atom?

What is the central concept in this sub-section?

**Electron Shells**

What is the Coulomb force?

What role does the Coulomb force play in an atom?

How is the binding energy of an electron related to the orbit of the electron?

If the binding energy of the electron is large, is the electron's orbit considered to be large or small?

What is meant by permitted orbits?

What are the central concepts of this sub-section?

**Window on Science 6-1. Quantum Mechanics: The World of the Very Small.**

What is quantum mechanics?

For what types of objects is quantum mechanics applicable?

How does quantum mechanics describe an electron?

What does quantum mechanics tell us about our ability to measure an object's position and motion?

What is the central concept in this Window on Science?

Consider the three sub-sections as a group, what are the fundamental ideas presented in this section?

**6-2** THE INTERACTION OF LIGHT AND MATTER

**The Excitation of Atoms**

What is meant by an energy level of an atom?

If an electron in an atom moves from one level to another level that has a greater energy, what has happened to the atom?

If an atom absorbs a photon, does the electron move to a higher or lower energy level in the atom?

In what two ways can an atom be excited?

It is said that an excited atom is unstable, what does this mean?

What is meant by the ground state of an atom?

If the electron is in the ground state can it produce a photon?

If an electron is in the ground state, can it absorb a photon?

What is the central concept of this sub-section?

### Radiation from a Heated Object

What is the difference between heat and temperature?

In the Kelvin system, what is the temperature of an object that contains no heat energy?

Why is the Kelvin scale useful?

How is black body radiation produced?

Black body radiation shows two important features related to temperature, what are they?

What does the symbol $\lambda_{max}$ represent?

In what way does $\lambda_{max}$ change when the temperature of an object changes?

If the temperature of a black body is doubled, what happens to the wavelength at which it radiates the most energy?

In what way does the energy emitted by a black body change when the temperature of an object changes?

If the temperature of a black body is doubled, what happens to the total amount of energy that it radiates per second from each square meter of its surface?

What can we learn about an object from the wavelength of its maximum intensity?

What is the central concept of this sub-section?

What are the fundamental ideas presented in this section?

### 6-3 STELLAR SPECTRA

### The Formation of a Spectrum and Atomic Spectra on pages 100 and 101

What is a spectrum?

What are the three types of spectra?

How is each type of spectra produced?

What is meant by the term transition as it applies to atomic spectra?

Write down each of Kirchhoff's three laws.

If an object produces an absorption spectrum, what do we know about that object?

What determines the absorption or emission lines present in a spectrum?

What is the central concept of this sub-section?

### The Balmer Thermometer

What atom produces the Balmer lines described in this section?

What transitions produce the Balmer lines in Hydrogen?

For most stars, are the Blamer lines in emission or absorption?

In what portion of the spectrum are Balmer lines found (X-ray, UV, visible, IR, or radio)?

At what temperatures are the Balmer lines the strongest?

Why are the Balmer lines weak at temperatures below 6,000 K?

Why are the Balmer lines weak at temperatures above 20,000 K?

What is the central concept of this sub-section?

### Spectral Classification

When was the first widely used spectral classification system developed and who was the principle scientist who developed it?

On what physical property of stars is the spectral sequence arranged?

List the spectral classes in order from hottest to coldest.

For which spectral class are the Balmer lines the strongest?

What is the central concept of this sub-section?

### The Doppler Effect

What is the Doppler Effect?

What is a blue shift?

What is a red shift?

What is meant by the term radial velocity?

Does the Doppler effect only apply to electromagnetic radiation?

What is the central concept of this sub-section?

What are the fundamental ideas presented in this section?

## KEY CONCEPTS

This is the most important chapter in the text. Almost everything we know about a star, gas cloud, and many solar system objects is determined from the object's spectrum, and nearly everything that goes on inside a star or gas cloud depends on the interaction of light and matter. The major points that need to be understood in this chapter are

1.   the necessity of physics in astronomy,
2.   the structure of the atom,
3.   the difference between heat and temperature,
4.   the use of the black body spectrum to determine temperature and energy output,
5.   the formation of spectral lines, and
6.   what can be learned from the spectrum of an object.

These concepts will be used throughout the remainder of the book.


## WEBSITES OF INTEREST

http://www.learner.org/teacherslab/science/light/color/spectra


## QUESTIONS ON CONCEPTS

1.   Graphs of the continuous spectra of five different stars are shown in the figure below. What can you determine about these stars from this figure?

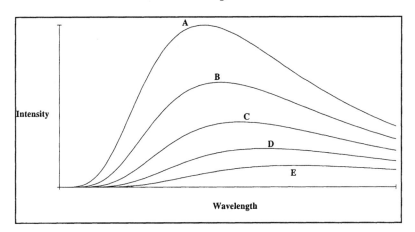

Solution:
The continuous spectra show a black body spectrum from which the temperature of the object can be determined. The temperature is determined from the wavelength at which the intensity is the greatest. If the peak in the intensity occurs at a short wavelength then the object is relatively hot, and if the peak occurs at a longer wavelength, then the object is relatively cool. In this figure, the peak in the intensity of Star A occurs at the shortest wavelength. Therefore, Star A is the hottest of the five stars. Since the peak in the intensity of Star E occurs at the longest wavelength, Star E must be the coolest.

Therefore the stars are of different temperatures, and arranged from lowest temperature to greatest temperature they are Star E, Star D, Star C, Star B, Satr A.

2. Arrange the transitions indicated below in order of energy released from smallest to largest.

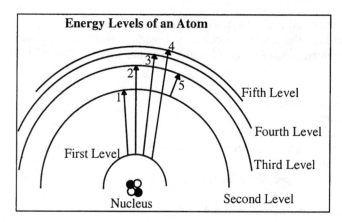

Solution:
The energy levels in an atom get closer and closer together as you move outward. Additionally, the energy released in a transition is equal to the difference between the two energy levels in the transition. Transitions 1, 2, 3 and 4 all occur from the first level, and the energies of the levels increase as you go outward. So of these four transitions, their order from smallest to greatest energy is Transition 1, Transition 2, Transition 3, and Transition 4. Transition 5 appears to have a smaller energy jump than even Transition 1. Since the energy levels do get closer as you move outward, the energy released in Transition 5 will be smaller than the energy released in Transition 1.

So the order of the transitions from smallest to largest energy released is Transition 5, Transition 1, Transition 2, Transition 3, Transition 4.

3. What is the order of the transitions in the problem above in terms of the wavelength of the photons produced by each transition?

Solution:
We know that high-energy photons have very short wavelengths and that low energy photons have very long wavelengths. This means that the transition above with the greatest energy, Transition 4, will produce a photon with the greatest energy and smallest wavelength. The transition with the lowest energy released, Transition 5, produces the photon with the longest wavelength.

So the order of the photons from shortest wavelength to longest wavelength is Transition 4, Transition 3, Transition 2, Transition 1, and Transition 5.

4. A star is found with very weak hydrogen lines, weak helium lines, and strong ionized helium lines, what is the approximate temperature of this star (see Figure 6-7)?

Solution:
Look carefully at Figure 6-7. Notice that the curve for hydrogen indicates that the hydrogen absorption lines will be relatively strong at temperatures around 10,000 K and weaker at temperatures both greater than and less than 10,000 K. That the hydrogen lines are weak indicates that the temperature must is either between 4,000 K and 8,000 K, or between 20,000 K and 40,000 K. The presence of weak helium lines implies that the temperature of the star is either between 10,000 K and 20,000 K, or greater than 30,000 K. Taken together the presence of weak hydrogen lines means that that temperature cannot be between 8,000K and 20,000 K, and the presence of the helium lines implies that the temperature cannot be less than 10,000K. So the temperature must be between 30,000 and 40,000 K The strong helium lines indicate that the temperature must be greater than 30,000 K, which confirms our determination of a temperature between 30,000 K and 40,000 K.

## WORKED EXAMPLES OF PROBLEMS REQUIRING MATHEMATICS

1. A planet is observed to produce a continuous spectrum with a maximum intensity at 20,000 nm. What is the surface temperature of this planet?

   Solution:
   We are given that the wavelength at which the object radiates its greatest intensity is 20,000 nm.
   Therefore we know that $\lambda_{max}$ = 20,000 nm.
   We want to find the temperature of the object, $T$.
   We know from Wien's Law, *By the Numbers 6-1* page 98 that

   $$\lambda_{max} = \frac{3,000,000}{T} = \frac{3.0 \times 10^6}{T}.$$

   Multiplying both sides by $T$ and cancel them on the right side, we find

   $$\lambda_{max} T = T \cdot \frac{3.0 \times 10^6}{T} = 3.0 \times 10^6.$$

   If we now divide both sides by $\lambda_{max}$, we have an equation that we can use to find the temperature, T.

   $$\frac{\lambda_{max} T}{\lambda_{max}} = T = \frac{3.0 \times 10^6}{\lambda_{max}}$$

   Substituting 20,000 nm in for $\lambda_{max}$ we have

   $$T = \frac{3.0 \times 10^6}{20,000} = \frac{3.0 \times 10^6}{2.0 \times 10^4} = 1.5 \times 10^2 \, \text{K}$$

   The temperature of the planet is $1.5 \times 10^2$ K or 150 K. This is approximately the temperature of Jupiter's cloud tops.

2. A chunk of ceramic is heated to a temperature of 7,000 K. At what wavelength will it produce the most energy, and what type of electromagnetic radiation is this?

   Solution:
   We are given the surface temperature as 7,000 K, so we know $T$ = 7,000 K.

   We will use Wien's Law, *By the Numbers 6-1* on page 98 to solve this problem.

   $$\lambda_{max} = \frac{3,000,000 \, \text{nm} \cdot \text{K}}{T}$$

   Substituting 7,000K for T, we find

   $$\lambda_{max} = \frac{3,000,000 \, \text{nm} \cdot \text{K}}{7,000 \, \text{K}} = 430 \, \text{nm}$$

   So the wavelength at which the most energy is emitted is 430 nm.
   This is in the visible portion of the spectrum and corresponds to blue-violet light.

3. Star A has a temperature of 3,200 K and star B has a temperature of 16,000 K.
   a) Which star radiates the most energy per second from each square meter of its surface?
   b) Compared to star A, how much more energy does star B emit per second from each square meter of its surface?

Solution to part a:
Both parts of this problem use the Stefan-Boltzmann Law described in *By the Numbers 6-1* on page 98.

In the equation $E = \sigma T^4$, $E$ represents the energy radiated per second by one square meter of surface for a black body. Notice that as the temperature, $T$ increases, so does the energy radiated per second per square meter of surface. Therefore, the hotter the object, the more energy it radiates per second. Consequently, Star B will radiate more energy per second from each square meter of surface than Star A does.

Solution to part b:
We again will use the Stefan-Boltzmann Law from *By the Numbers 6-1* on page 98. First, we write down what we know.
Temperature of Star A = $T_A$ = 3,200 K = $3.2 \times 10^3$ K.
Temperature of Star B = $T_B$ = 16,000 K = $1.6 \times 10^4$ K.

We are asked to find how much more energy Star B produces than Star A. We really want to know what the ratio of the energy produced by Star B is compared to Star A. So we want to calculate $\dfrac{E_B}{E_A}$.

$$E_B = \sigma T_B^4 = \sigma \left(1.6 \times 10^4 \, \text{K}\right)^4 \quad \text{and} \quad E_A = \sigma T_A^4 = \sigma \left(3.2 \times 10^3 \, \text{K}\right)^4$$

So

$$\frac{E_B}{E_A} = \frac{\sigma \left(1.6 \times 10^4 \, K\right)^4}{\sigma \left(3.2 \times 10^3 \, K\right)^4}$$

The $\sigma$ in the numerator cancels the one in the denominator, and

$$\frac{E_B}{E_A} = \frac{\sigma \left(1.6 \times 10^4 \, \text{K}\right)^4}{\sigma \left(3.2 \times 10^3 \, \text{K}\right)^4} = \left(\frac{1.6 \times 10^4}{3.2 \times 10^3}\right)^4 = (5)^4 = 625$$

Therefore Star B radiates 625 times more energy per second from each square meter of surface than does star A.

4. The Lyman delta line has a wavelength of 95.0 nm in the laboratory. In the spectrum of a star, the Lyman delta line appears at a wavelength of 94.9 nm.
   a) Is this star moving toward or away from Earth?
   b) At what speed is it moving relative to Earth?

Solution to part a:
This problem concerns the Doppler shift of a spectral line discussed in *By the Numbers 6-2* on page 107. To determine whether the object is moving toward or away from us, we need to determine if the spectral feature shows a red shift, which would imply that the star is moving away from us, or if the spectral feature shows a blue shift, which indicates that the star is moving toward Earth. We know that in the laboratory, the Lyman delta line in the hydrogen spectrum appears at 95.0 nm. We are told that the star emits a spectrum such that the Lyman delta line appears at 94.9 nm. Note that the line in the star's spectrum appears at a shorter wavelength than it appears at in the laboratory spectrum.

This means that the star's spectrum is blue shifted, and hence the star is moving toward us.

Solution to part b:
We know:
the wavelength of the spectral line in the laboratory = $\lambda_o$ = 95.0 nm
the wavelength of the line as observed in the star = $\lambda$ = 94.9 nm, and
the speed of light = $c$ = $3.0 \times 10^5$ km/sec

We can calculate $\Delta\lambda = \lambda_o - \lambda$ = 95.0 nm – 94.9 nm = 0.1 nm

We also know that $\dfrac{\Delta\lambda}{\lambda_o} = \dfrac{v}{c}$

We know everything except $v$ and that is what we want to find, so lets rewrite the equation so we can solve for v. To do this we multiply both sides by $c$. The $c$'s will cancel on the right side of the equation.

$$c \cdot \frac{\Delta\lambda}{\lambda_o} = c \cdot \frac{v}{c}$$

$$c \cdot \frac{\Delta\lambda}{\lambda_o} = v$$

Now simply turn the equation around and enter the known quantities.

$$v = c \cdot \frac{\Delta\lambda}{\lambda_o} = 3.0 \times 10^5 \text{ km/sec} \cdot \frac{0.1 \text{ nm}}{95.0 \text{ nm}} = \frac{3.0 \times 10^4}{95.0} \text{ km/sec} = 3.16 \times 10^2 \text{ km/sec} = 316 \text{ km/s}$$

So the star's radial velocity is 316 km/sec toward Earth.

5.  If a galaxy is moving away from Earth at a speed of 500 km/sec, at what wavelength will the Lyman delta line appear in its spectrum (see problem 4 above)?

Solution:
We know:
the radial velocity of the galaxy = $v$ = 500 km/sec,
the wavelength of the line in the laboratory = $\lambda_o$ = 95.0, and
the speed of light = $c$ = $3.0 \times 10^5$ km/sec.

We can find the wavelength the line will appear at by solving the Doppler equation of $\Delta\lambda$ and then solving for $\lambda$.

$$\frac{\Delta\lambda}{\lambda_o} = \frac{v}{c}$$

We want to rearrange this equation so that we have $\Delta\lambda$ equal to things we know. So we need to multiply both sides by $\lambda_o$.

$$\lambda_o \cdot \frac{\Delta\lambda}{\lambda_o} = \lambda_o \cdot \frac{v}{c}$$

The $\lambda_0$'s on the left side of the equation cancel, and we are left with,

$$\Delta\lambda = \lambda_0 \cdot \frac{v}{c} = 95.0 \text{ nm} \cdot \frac{500 \text{ km/sec}}{3.0 \times 10^5 \text{ km/sec}} = 95.0 \text{ nm} \cdot \left(1.67 \times 10^{-3}\right) = 0.16 \text{ nm}$$

So we know $\Delta\lambda = 0.16$ nm and we know that the galaxy is moving away from us. Since the galaxy is moving away from us, the line will appear at a longer wavelength than it does in the laboratory. Therefore the line in the galaxy's spectrum is at

$$\lambda_{galaxy} = \lambda_0 + \Delta\lambda = 95.0 \text{ nm} + 0.16 \text{ nm} = 95.16 \text{ nm}.$$

The Lyman delta line will appear at a wavelength of 95.16 nm in the galaxy's spectrum.

## MORE MATH RELATED PROBLEMS
(Answers at the end of the chapter)

1. The human body temperature is approximately $38^\circ$ C or 311 K. At what wavelength does the human body radiate the most energy?

2. The star Vega radiates most strongly at 300 nm, what is the approximate temperature of Vega?

3. Below is a table of objects with various temperatures. For some the temperature is known and you need to determine the wavelength at which the maximum energy is emitted. For others the wavelength at which the maximum energy is emitted is known and you need to determine the temperature. Complete the table.

| Object | Temperature | $\lambda_{max}$ |
|---|---|---|
| Pluto | | 60,000 nm |
| Very hot star | | 50 nm |
| Orange star | 4,300 K | |
| Venus | 600 K | |
| White star | | 231 nm |
| Jupiter | 143 K | |
| Yellow-white star | | 400 nm |
| Chair in class | 293 K | |
| Center of Sun | | 0.2 nm |
| Ice cube | 273 | |

4. Venus has a surface temperature of 600 K and the sun has a surface temperature of approximately 6,000 K. How much more energy does the sun radiate from each square meter of its surface compared to that of Venus?

5.  Given below is a table of objects with different temperatures.  Determine which object radiates the most energy per square meter, and how much more energy it radiates per square meter.  The first one is completed as an example.

| Object A | Object B | Radiates Most Energy per $m^2$ | Ratio of Energy Radiated |
|---|---|---|---|
| 300 K | 400 K | Object B | 3.2 times |
| 100 K | 50 K | | |
| 7,000 K | 21,000 K | | |
| 64,000 K | 4,000 K | | |
| 27,000 K | 9,000 K | | |
| 273 K | 373 K | | |
| 5,800 K | 3,000 K | | |
| 5,800 K | 70,000 K | | |

6.  A star is found with the Lyman β line of hydrogen at a wavelength of 103.5 nm.  In the laboratory the Lyman β line of hydrogen appears at a wavelength of 102.6 nm.
    a. Is this star moving toward or away from us?
    b. What is the radial velocity of this star?

7.  A satellite is traveling at 100 km/sec away from Earth and sending data back to Earth at a wavelength of 1 m.  What will the wavelength of this signal be here at Earth?

8.  Below is a table of wavelengths and radial velocities.  Complete the table.

| Object | Laboratory Wavelength | Observed Wavelength | Approaching or Receding? | Radial Velocity |
|---|---|---|---|---|
| 1 | 656.3 nm | 656.1 nm | | |
| 2 | 121.5 nm | 121.9 nm | | |
| 3 | 656.3 nm | 656.7 nm | | |
| 4 | 388.9 nm | | Approaching | 617 km/sec |
| 5 | 486.1 nm | | Receding | 741 km/sec |
| 6 | 121.5 nm | | Receding | 3,210 km/sec |
| 7 | | 391.5 nm | Receding | 2,006 km/sec |
| 8 | | 121.2 nm | Approaching | 741 km/sec |
| 9 | | 1865.6 nm | Approaching | 991 km/sec |
| 10 | 434.0 nm | 434.0 nm | | |

**PRACTICE TEST**
**Multiple Choice Questions**

1.  Absolute zero is
    a. zero degrees Celsius.
    b. the temperature at which atoms have no remaining energy from which heat can be extracted.
    c. the temperature at which water freezes.
    d. both a and c
    e. none of the above

2.  A neutral atom consists of
    a. one proton and one neutron.
    b. only protons and neutrons.
    c. only electrons.
    d. the same number of protons as electrons.
    e. at least two atoms bound together by the Coulomb force.

3.  Ionization
    a.  is the process of removing an electron from a neutral atom.
    b.  is the formation of an emission line by an excited atom.
    c.  causes a Doppler shift in spectral lines.
    d.  produces a nucleus that contains more protons than neutrons.
    e.  increases the number of electrons that an atom has.

4.  Atom A has 3 protons and 3 neutrons, while atom B has 3 protons and 2 neutrons.  Which of the
    following is/are true.
    a.  Atom A and Atom B are ions of the same element.
    b.  Atom A and Atom B are isotopes of the same element.
    c.  Atom A and Atom B produce identical spectra.
    d.  All of the above
    e.  None of the above

5.  An atom can become excited
    a.  if it emits a photon.
    b.  if it collides with another atom or electron.
    c.  if it absorbs a photon.
    d.  a and b
    e.  b and c

6.  A plot of the continuous spectra of five different stars are shown in the figure below.  Based on these
    spectra, which of the stars is the hottest?

    a.  Star A
    b.  Star B
    c.  Star C
    d.  Star D
    e.  Star E

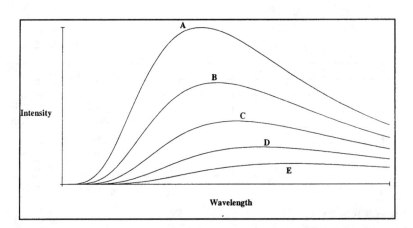

7.  In the diagram below, which of the transitions would absorb a photon with the greatest energy.
    a.  Transition 1
    b.  Transition 2
    c.  Transition 3
    d.  Transition 4
    e.  Transition 5

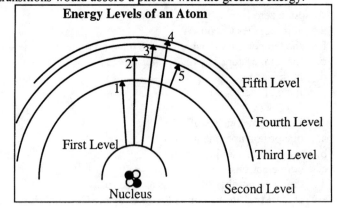

8. An atom that is excited
   a. is also ionized.
   b. is an isotope.
   c. has had its electron moved to the lowest energy level.
   d. can emit a photon when the electron moves to a lower energy level.
   e. can emit a photon when the electron moves to a higher energy level.

9. The radiation emitted from a star has a maximum intensity at a wavelength of 10,000 nm. What is the temperature of this star?
   a. 300 K
   b. 100 K
   c. 900,000,000 K
   d. 90,000 K
   e. 10,000 K

10. The diagram to the right illustrates a light source, a gas cloud, and three different lines of sight. Along which line of sight would an observer see an emission spectrum?
   a. 1
   b. 2
   c. 3
   d. 2 and 3
   e. none of them

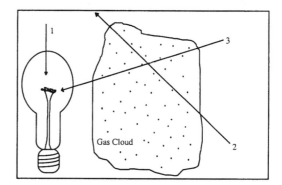

11. At what wavelength would a star radiate the greatest amount of energy if the star has a surface temperature of 5,000 K?
   a. 60 nm
   b. 600 nm
   c. 500 nm
   d. $1.5 \times 10^{10}$ nm
   e. 180 nm

12. One star has a temperature of 12,000 K and another star has a temperature of 4,000 K Compared to the cooler star, how much more energy per second will the hotter star radiate from each square meter of its surface?
   a. 3 times
   b. 9 times
   c. 81 times
   d. $9 \times 10^6$ times
   e. 16,000 times

13. The $H_\delta$ line has a wavelength of 410.2 nm when observed in the laboratory. If the $H_\delta$ line appears in a stars spectrum at 410.7 nm, what is the radial velocity of the star?
   a. 366 km/sec away from the observer.
   b. 366 km/sec toward the observer.
   c. $1.5 \times 10^8$ m/sec away from the observer.
   d. $1.5 \times 10^8$ m/sec toward the observer.
   e. The radial velocity of the star cannot be determined from this information.

## Fill in the Blank Questions

14. _____ is a set of rules that describe how atoms and subatomic particles behave.

15. The process of removing an electron from an atom is known as _____.

16. An atom can be excited by _____ or _____

17. If one star has a temperature of 6,000 K and another star has a temperature of 30,000 K, how much more energy per second will the hotter star radiate from each square meter of its surface? _____

## True-False Questions

18. If an object is heated to twice its current temperature, it will radiate 2 times more energy per square meter of its surface than it currently does.

19. The faster an object moves toward you, the greater the Doppler shift in that object's spectrum.

20. Kirchhoff's laws state that a low-density gas excited to emit light at specific wavelengths produces an absorption spectrum.

21. The lowest energy state of an electron in an atom is known as the ground state.

22. The Coulomb force binds the neutrons and protons together to form the nucleus of the atom.

## ADDITIONAL READING

Feinberg, G. *What is the World Made Of?* New York: Doubleday (Anchor), 1977.

Kaler, James B. "Beyond the Rainbow." *Astronomy 28* (Sept. 2000), p. 38.

Kaler, James B. *Stars and Their Spectra.* New York: Cambridge University Press, 1989.

Walker, Jearl. "The Amateur Scientist: The Spectra of Streetlights Illuminate Basic Principles of Quantum Mechanics." *Scientific American 250* (Jan. 1984), p. 138.

## ANSWERS TO MORE MATH RELATED PROBLEMS

1. 9,646 nm (9.65 μm) in the infrared.
2. 10,000 K
3. Answers appear in **bold** type

| Object | Temperature | $\lambda_{max}$ |
|---|---|---|
| Pluto | **50 K** | 60,000 nm |
| Very hot star | **60,000 K** | 50 nm |
| Orange star | 4,300 K | **698 nm** |
| Venus | 600 K | **5,000 nm** |
| White star | **13,000 K** | 231 nm |
| Jupiter | 143 K | **21,000 nm** |
| Yellow-white star | **7,500 K** | 400 nm |
| Chair in class | 293 K | **10,240 nm** |
| Center of Sun | $1.5 \times 10^7$ K | 0.2 nm |
| Ice cube | 273 | **11,000 nm** |

4. $10^4$ times = 10,000 times.

5.

| Object A | Object B | Radiates Most Energy per m$^2$ | Ratio of Energy Radiated |
|---|---|---|---|
| 300 K | 400 K | Object B | 3.2 times |
| 100 K | 50 K | Object A | 16 times |
| 7,000 K | 28,000 K | Object B | 256 times |
| 64,000 K | 4,000 K | Object A | 65,500 times |
| 27,000 K | 9,000 K | Object A | 81 times |
| 273 K | 373 K | Object B | 3.5 times |
| 5,800 K | 3,000 K | Object A | 14 times |
| 5,800 K | 70,000 K | Object B | 21,000 times |

6. a. Away from us.
   b. 2,632 km/sec
7. 1.00033 m
8.

| Object | Laboratory Wavelength | Observed Wavelength | Approaching or Receding? | Radial Velocity |
|---|---|---|---|---|
| 1 | 656.3 nm | 656.1 nm | Approaching | 91 km/sec |
| 2 | 121.5 nm | 121.9 nm | Receding | 988 km/sec |
| 3 | 656.3 nm | 656.7 nm | Receding | 183 km/sec |
| 4 | 388.9 nm | 388.1 nm | Approaching | 617 km/sec |
| 5 | 486.1 nm | 487.3 nm | Receding | 741 km/sec |
| 6 | 121.5 nm | 122.8 nm | Receding | 3,210 km/sec |
| 7 | 388.9 nm | 391.5 nm | Receding | 2,006 km/sec |
| 8 | 121.5 nm | 121.2 nm | Approaching | 741 km/sec |
| 9 | 1875.1 nm | 1865.6 nm | Approaching | 991 km/sec |
| 10 | 434.0 nm | 434.0 nm | Neither | 0 km/sec |

## ANSWERS TO PRACTICE TEST

1. b
2. d
3. a
4. b
5. e
6. a
7. d
8. d
9. a

10. b
11. b
12. c
13. a
14. Quantum mechanics
15. ionization
16. the absorption of a photon or by collisions with other atoms or particles

17. 625 times
18. F
19. T
20. F
21. T
22. F

# CHAPTER 7

# THE SUN - OUR STAR

## UNDERSTANDING THE CONCEPTS

**7-1    The Solar Atmosphere**
**Heat Flow in the Sun**
What is the temperature of the sun's surface?
What is the source of energy in the sun?
What is the central concept of this sub-section?
**The Photosphere**
What is the photosphere?
What type of spectrum does the photosphere of the sun produce?
What is granulation?
What do granules on the sun's surface look like?
What is the central concept of this sub-section?
**The Chromosphere**
Where is the chromosphere of the sun located?
When is the chromosphere visible to the unaided eye?
How does the temperature of the chromosphere change as a function of height above the photosphere?
What is a filtergram and why are they important in studying the sun?
What are filaments on the sun?
What are spicules?
What is the central concept of this sub-section?
**The Solar Corona**
What features in the sun's photosphere are linked to the corona and chromosphere by the magnetic field?
What is the temperature of much of the coronal gas?
What mechanism is believed to heat the corona to higher temperatures than the photosphere?
What is the solar wind?
What fraction of the sun is lost each year in the solar wind?
What is the central concept of this sub-section?
**Helioseismology**
What is helioseismology?
Helioseismology has allowed astronomers to measure three quantities in the solar interior, what are they?
What is the central concept of this sub-section?

What are the fundamental ideas presented in this section?

**7-2    Solar Activity**
**Sunspots**
What is the typical diameter of a sunspot?
How long does a typical sunspot group exist?
What is the typical temperature of a sunspot?
What is the duration of the sunspot cycle?
What is the Maunder minimum?
What is the central concept of this sub-section?
**Active Regions and Sunspots**
What effect can we use to measure the strength of the magnetic field in sunspots?
What is the central concept of this sub-section?

**The Sun's Magnetic Cycle**

Which rotates faster in the sun, the material near the poles or the material near the equator?

What is the dynamo effect?

Describe the Babcock model.

What is the central concept of this sub-section?

**Chromospheric and Coronal Activity and Magnetic Solar Phenomenon on pages 124 and 125**

What are two important points about solar activity?

How are aurora produced?

What is a prominence?

What is a coronal mass ejection?

Where do solar flares occur?

What is the central concept of this sub-section?

**Window on Science 7-1. Building Confidence by Confirmation and Consolidation**

What is meant by confirmation in science?

How does confirmation aid in consolidation?

What is the central concept of this sub-section?

What are the fundamental ideas presented in this section?

**7-3     Nuclear Fusion in the Sun**

**Nuclear Binding Energy**

What is the name of the force that binds protons and neutrons together to form nuclei?

What type of nuclear reaction occurs in the cores of most stars?

Why is energy released in a nuclear fusion reaction?

What is the central concept of this sub-section?

**Hydrogen Fusion**

How many reactions must occur inside the sun each second to keep gravity from collapsing it?

What is the Coulomb barrier?

What is the proton-proton chain?

What nuclear reaction is occurring in the center of the sun?

What is the minimum temperature required for the proton-proton chain to occur?

What three forms does the energy generated in the proton-proton chain take?

What is the central concept of this sub-section?

**The Solar Neutrino Problem**

What is a neutrino?

Are neutrinos harmful to the human body?

Why were the results from the Davis solar neutrino experiment startling to astronomers?

How has the results of the Davis experiment changed our understanding of the neutrino?

What is the central concept of this sub-section?

**Window on Science 7-2. Avoiding Hasty Judgments: Scientific Faith**

What do we call a theory that has been tested many time and in which scientists have a great deal of faith?

Why don't scientists immediately abandon a theory when evidence is reported that contradicts the theory?

What is the central concept of this sub-section?

What are the fundamental ideas presented in this section?

**KEY CONCEPTS**

The major emphases of this chapter are the structure and phenomena of the observable layers of the sun and the mechanism by which the sun produces its energy. This discussion serves as an introduction to the study of the observable properties of stars in general.

The first section describes the various observable layers and the methods used to determine the temperature and density of these layers. The photosphere has been studied primarily with optical spectroscopy and shows an absorption spectrum. This layer is the visible surface of the sun, has a temperature of about 5800 K and is about 500 km thick. The photosphere contains granulation, which indicates the importance of convection within the sun. We also find sunspots on the surface, which are discussed in greater detail in the next section.

The chromosphere is hotter and less dense than the photosphere. It produces an emission spectrum that shows that many of the atoms in the chromosphere are ionized. The chromosphere is studied with filtergrams and also during eclipses.

The corona is the outer most layer of the solar atmosphere and is extremely hot at 1 million K. It is also very low density. The corona has been studied during total solar eclipses and using a coronagraph. The heat flow from the cooler photosphere to the hotter chromosphere and corona has presented a problem for astronomers for a long time. A possible solution suggests that magnetic fields originating in the photosphere and extending into the chromosphere and corona may whip around and agitate these regions causing an increase in their temperatures.

Helioseismology is the study of vibrations in the sun and their effects on the photosphere. From helioseismology, astronomers are able to map the temperature, density, and rate of rotation inside the sun. This is one of the few techniques that lets us receive information about the interior of the sun.

The second section emphasizes the magnetic field of the sun and the observed variations in the magnetic field. Most of the activity of the sun is a result of the magnetic field and the sun's rotation. Sunspots are the most obvious and best known active regions on the sun. Sunspots are cooler than the surrounding photosphere and are associated with magnetic loops extending from the photosphere. The magnetic field suppresses the flow of hot gas and energy in these regions. The Maunder butterfly diagram shows us that the spots follow a definite eleven-year cycle in their location on the sun's surface.

The Zeeman effect is essential in studying the active regions of the sun. The magnetic fields cause the emission or absorption lines in a spectrum to split. The amount of splitting is directly proportional to the strength of the magnetic field. This gives us a way to measure the strength of the magnetic field at different locations on the sun. We find that sunspots and other active regions show stronger magnetic field strengths than the rest of the surface.

The final section describes nuclear fusion and the production of energy in the core of the sun through the proton-proton chain. One of the products of the proton-proton chain is neutrinos. The solar neutrino problem is an important issue that has been around for a long time. Its solution is important to our understanding of neutrinos and stellar evolution, as well as the workings of our sun's interior. The attempt to understand the rate at which neutrinos are produced changed our understanding of neutrinos. Our confidence in our theories about how energy is produced in the sun and stars lead us to ask questions about the neutrino itself. We now believe that the neutrino has a very small mass and comes in three different types.

In all three sections, relationships between phenomena found on the sun and observations of other stars are discussed to indicate that the sun is a normal star. One of the major concepts of this chapter is that astronomy is more than identifying and classifying phenomena, it also develops models and theories to help us understand how these phenomena are produced and what that tells us about the physics of the star's atmosphere and interior.

## QUESTIONS ON CONCEPTS

1.  How are aurora related to solar activity?

    Solution:
    The light of aurorae is produced as charged particles collide with molecules in Earth's atmosphere. These charged particles are emitted by the sun and get trapped in Earth's magnetic field, which funnels them into the atmosphere near the magnetic poles. The sun always emits a relatively small number of particles in the solar wind. During times of extreme solar activity, a much larger number of charged particles are dumped into the solar wind. With more charged particles flowing through the solar system, a larger number of them are collected by the magnetic field and funneled into Earth's atmosphere. With a larger number of charged particles bombarding the molecules of the atmosphere, more light is produced in the aurora during times of intense solar activity. In a way, the aurora occurs because of an interaction between the sun's magnetic field and Earth's magnetic field. When the sun's field is strong and rapidly changing, it produces larger amounts of charged particles, some of which get trapped in Earth's magnetic field.

2.  The diagram at the right shows a plot of the temperature of the sun as a function of distance above the bottom of the photosphere. At what distance above the bottom of the photosphere does the temperature of the sun change the most rapidly with distance?

    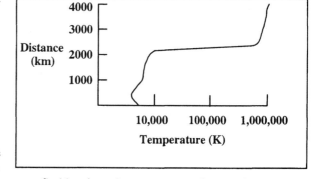

    Solution:
    First look at how the axes of the graph are labeled, the temperature is plotted along the bottom and distance from the bottom of the photosphere is plotted vertically. What we want to find is where the temperature changes the most over the smallest distance. At a distance just greater than 2000 km, the temperature is about 10,000 K and at a distance of less than 2500 km the temperature has risen to nearly 1,000,000 K. So between 2000 km and 2500 km, the temperature changes from 10,000 K to 1,000,000 K.

3.  As the moon covers the solar disk during a solar eclipse, a flash spectrum of the sun's chromosphere can be recorded. This flash spectrum reveals an emission spectrum and provides information on the properties of the chromosphere. As the moon moves from the inner chromosphere to the outer chromosphere, the spectral lines present in the flash spectrum change. What is going on in the chromosphere that produces the changes in the flash spectrum?

    Solution:
    The emission lines present in any spectrum depend on the temperature of the gas. From Figure 7-3 in the text, we can see that the temperature in the chromosphere varies with distance from the photosphere. At the base of the chromosphere the temperature is about 4500 K and it increases to about 10,000 K at the top of the chromosphere. In chapter 6 we learned that at 10,000 K the lines of hydrogen are be very strong, while at 4500 K the lines of hydrogen are quite weak, but the lines of calcium are strong. So we expect the flash spectrum to change as we record the spectrum of the

different regions of the chromosphere with the flash spectrum. The use of flash spectra allows us to understand the changes in the temperature and density of the chromosphere.

## WORKED EXAMPLES OF PROBLEMS REQUIRING MATHEMATICS

1.  If a star identical to our sun were located 2 ly from us, what would its angular diameter be?

    Solution:
    We know:
    a star identical to our sun will have a diameter equal to our sun's = $1.4 \times 10^6$ km,
    1 light-year = $9.5 \times 10^{12}$ km, and
    the distance to the star is 2 ly = 2 ($9.5 \times 10^{12}$ km) = $1.9 \times 10^{13}$ km.
    We use the small angle formula from *By the Numbers 3-1* on page 32.

    $$\frac{angular\ diameter}{206,265"} = \frac{linear\ diameter}{distance}$$

    If we multiply both sides by 206,265" we can solve for the angular diameter.

    $$(206,265")\cdot\frac{angular\ diameter}{206,265"} = (206,265")\cdot\frac{linear\ diameter}{distance}$$

    $$angular\ diameter = (206,265")\cdot\frac{1.4\times10^6\ km}{1.9\times10^{13}\ km} = 0.015\ seconds\ of\ arc.$$

    So a star identical to our sun located 2 ly from Earth would have an angular diameter of only 0.015 second of arc.

2.  In a large nuclear reaction 0.5 kg of hydrogen is converted into energy. How much energy is produced?

    Solution:
    We know how much mass was converted into energy = 0.5 kg.
    We can use Einstein's energy equation to solve for the energy produced.

    $$E = mc^2 = (0.5\,kg)\cdot(3.0\times10^8\ m/s)^2 = (0.5\,kg)\cdot(9.0\times10^{16}\ m/s) = 4.5\times10^{16}\ J.$$

    So $4.5\times10^{16}$ Joules of energy are produced when 0.5 kg of mass is converted to energy.

3.  One mega-ton of TNT releases $4.0\times10^{15}$ J of energy. How many mega-tons of TNT would be needed to produce as much energy as produced in converting 0.5 kg of hydrogen into energy?

    Solution:
    From the preceding problem we see that when 0.5 kg of hydrogen are completely turned into energy it produces $4.5\times10^{16}$ J. Since each mega-ton of TNT releases $4.0\times0^{15}$ J we find that

    $$TNT\ equivalent = \frac{energy\ released\ by\ hydrogen}{energy\ released\ by\ TNT} = \frac{4.5\times10^{16}\ J}{4.0\times10^{15}\ J} = 11\ megatons\ of\ TNT$$

    It takes 11 mega-tons of TNT to produce as much energy as is released when 0.5 kg of material is converted completely into energy.

4.  In a nuclear reaction, 4 kg of hydrogen are used to produce helium and energy.
    a.  How much helium is created in this reaction if all of the hydrogen is used?
    b.  How much energy is produced in the reaction?

Solution to part a:
Read *By the Number 7-1* on page 126. In each reaction 4 hydrogen nuclei with a combined mass of $6.693 \times 10^{-27}$ kg are converted into one helium nucleus with a mass of $6.645 \times 10^{-27}$ kg.
For each reaction the ratio of the mass of helium produced to the mass of hydrogen used is

$$\frac{mass\ of\ helium\ produced}{mass\ of\ hydrogen\ used} = \frac{6.645x10^{-27}\ kg}{6.693x10^{-27}\ kg} = 0.9928$$

This ratio is the same regardless of the initial amount of hydrogen. So for every kg of hydrogen used, the reaction produces 0.9928 kg of helium.

$$mass\ of\ helium\ produced = 0.9928 \cdot (mass\ of\ hydrogen\ used) = 0.9928 \cdot 4\ kg = 3.971\ kg$$

Therefore, the total amount of helium produced is 3.971 kg.

Solution to part b:
From part a we know that 4 kg of hydrogen produces 3.971 kg of helium. The amount of mass converted into energy then is equal to the difference between the mass of hydrogen and the mass of helium. The mass converted to energy, *m*, is given by

$$m = mass\ of\ hydrogen - mass\ of\ helium = 4\ kg - 3.971\ kg = 0.0287\ kg.$$

The energy produced is given by Einstein's energy equation

$$E = mc^2 = (0.0287\ kg) \cdot \left(3.0 \times 10^8\ \tfrac{m}{s}\right)^2 = (0.0287\ kg) \cdot \left(9.0 \times 10^{16}\ \tfrac{m^2}{s^2}\right) = 2.58 \times 10^{15}\ J.$$

4 kg of hydrogen will be converted into 3.971 kg of helium and $2.58 \times 10^{15}$ J of energy will be produced.

**MORE MATH RELATED PROBLEMS**
    (Answers at the end of the chapter)

1.  How much energy is released when 0.1 kg of mass is completely converted into energy?

2.  What mass of helium is produced when 10 kg of hydrogen are converted to helium by nuclear fusion?

3.  How much energy is produced when 10 kg of hydrogen are converted into helium by nuclear fusion?

4.  A solar flare releases $10^{25}$ J of energy.
    a.  What equivalent mass would need to be converted to energy to produce this much energy in a nuclear fusion reaction?
    b.  How much hydrogen would you need to start with to create this much energy by nuclear fusion?

5.  Assume that a typical starspot is equal to 10 times the diameter of Earth and that the star is 5 ly away. What is the angular diameter of this starspot?

6. Sirius produces $9.0 \times 10^{27}$ J/s.
   a. How much mass must Sirius convert to energy each second to produce this much energy?
   b. How much hydrogen does Sirius convert to helium and energy each second?
   c. How much helium does Sirius produce each second?

**PRACTICE TEST**
**Multiple Choice Questions**

1. Which part of the sun does helioseismology help us study?
   a. Corona.
   b. Chromosphere.
   c. Solar wind.
   d. Photosphere.
   e. Interior.

2. The diagram at the right shows a plot of the temperature of the sun as a function of distance above the bottom of the photosphere. What is the approximate temperature of the sun at a distance of 3,000 km above the bottom of the photosphere?
   a. 900 K
   b. 9,000 K
   c. 90,000 K
   d. 900,000 K
   e. 9,000,000 K

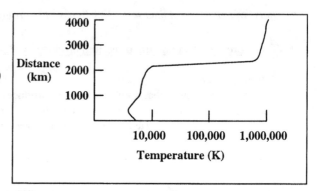

3. The solution to the neutrino problem indicates that
   I.   the neutrino has a small mass.
   II.  neutrinos exist in three different varieties.
   III. our understanding of how the sun creates energy needs to be significantly modified.
   IV.  the sun is significantly older than we had previously thought.

   a. I & II
   b. I & III
   c. I, II, & III
   d. I, II, & IV
   e. I, II, III, & IV

4. Which of the following is true for the proton-proton chain?
   a. In the proton-proton chain two protons are combined to form one helium nucleus.
   b. In the proton-proton chain 4 hydrogen nuclei are combined to form one helium nucleus.
   c. Solar flares are produced by the proton-proton chain
   d. The proton-proton chain requires that the temperature be no greater than 1,000,000 K
   e. The proton-proton chain is a nuclear fission reaction

5. The _____ of the sun is composed of ionized gas and produces a continuous spectrum with a superimposed emission spectrum associated with a temperature of about 1,000,000K.
   a. photosphere.
   b. chromosphere.
   c. corona.
   d. interior.
   e. none of the above.

6. The _____ occurred between 1645 and 1715 when very few sunspots occurred.
   a. Zeeman effect
   b. Maunder minimum
   c. umbra
   d. great San Francisco earthquake
   e. magnetic carpet

7. Regions on the sun where the Zeeman effect is observed to be strong are regions
   a. where the magnetic field is strong.
   b. where the temperature is greater than the average surface temperature.
   c. where granules are smaller than normal.
   d. associated weak magnetic fields.
   e. containing very few sunspots.

8. Sunspots
   I.   are hotter than the regions surrounding them.
   II.  are associated with strong magnetic fields.
   III. are located near the poles of the sun.
   IV.  show an 11 year cycle in the number that occur and in their location on the sun.

   a. I & II
   b. I & IV
   c. II & III
   d. II & IV
   e. I, II, & IV

9. The Coulomb barrier
   a. holds the neutron and proton together in the nucleus.
   b. is the force of repulsion between two nuclei when the two nuclei have opposite charge.
   c. is the force of repulsion between two nuclei because both have positive charge.
   d. keeps the corona from ejecting charged particles into the solar wind.
   e. keeps the gas in the photosphere from evaporating into space.

10. How much energy is produced when $5.0 \times 10^9$ kg is completely converted to energy in nuclear fusion.
    a. 17 J
    b. $5.6 \times 10^{-8}$ J
    c. $1.5 \times 10^{18}$ J
    d. $4.5 \times 10^9$ J
    e. $4.5 \times 10^{26}$ J

**Fill in the Blank Questions**

11. The _____ of the sun produces an emission spectrum associated with a gas at a temperature that ranges from about 4,500 K up to 10,000 K.

12. _____ are long dark string-like complexes in the sun's chromosphere.

13. The _____ model explains the magnetic cycle as a progressive tangling of the solar magnetic field.

14. The dynamo effect is believed to produce the _____ of the sun.

15. _____ occur when charged particles from the sun are trapped in Earth's magnetic field and collide with molecules in Earth's atmosphere near the magnetic poles.

**True-False Questions**

16. The neutrino problem has changed our understanding of the neutrino.

17. The nuclear binding energy of iron is stronger than that for any other nuclei.

18. The dynamo effect is responsible for producing the visible light coming from the sun.

19. The corona of the sun has a higher density than the photosphere.

20. Sunspots originate from magnetically disturbed regions.


## ADDITIONAL READING

Bartusiak, Marcia "Underground Astronomer." *Astronomy 28* (Jan. 2000), p. 64.

Burch, James L., "The Fury of Space Storms." *Scientific American 284*, (April 2001), p. 86.

Dwivedi, Bhala N. and Kenneth J. H. Phillips "The Paradox of the Sun's Hot Corona." *Scientific American 284*, (June 2001), p. 40.

Hayden, Thomas "Curtain Call." *Astronomy 28* (Jan. 2000), p. 44.

Human, Katy "When the Solar Wind Blows" *Astronomy 28* (Jan. 2000), p. 56.

Ortega, Tony "Quaking Sun" *Astronomy 28*, (Jan. 2000), p. 60.

## ANSWERS TO MORE MATH RELATED PROBLEMS
1. $9.0 \times 10^{15}$ J
2. 9.928 kg
3. $6.48 \times 10^{15}$ J
4. a. $1.11 \times 10^{8}$ kg
   b. $1.55 \times 10^{10}$ kg
5. $5.6 \times 10^{-5}$ seconds of arc, about a factor of 200 better than our current imaging capabilities.
6. a. $1.0 \times 10^{11}$ kg
   b. $1.389 \times 10^{13}$ kg
   c. $1.379 \times 10^{13}$ kg


## ANSWERS TO PRACTICE TEST
| | | |
|---|---|---|
| 1. e | 8. d | 15. Aurora |
| 2. d | 9. c | 16. T |
| 3. a | 10. e | 17. T |
| 4. b | 11. chromosphere | 18. F |
| 5. c | 12. Filaments | 19. F |
| 6. b | 13. Babcock | 20. T |
| 7. a | 14. magnetic field | |

# CHAPTER 8

# THE FAMILY OF STARS

## UNDERSTANDING THE CONCEPTS

**8-1    Measuring the Distances to Stars**
**The Surveyor's Method**
　　　This section is mostly introductory.
**The Astronomer's Method**
　　　Define Parallax.
　　　What is stellar parallax?
　　　What is the greatest stellar parallax angle and what is the distance to that star?
　　　What is a parsec?
　　　What is the greatest distance at which we can measure stellar parallaxes reliably?
　　　What is the central concept of this sub-section?

What are the fundamental ideas presented in this section?

**8-2    Intrinsic Brightness**
**Brightness and Distance**
　　　What is flux?
　　　How does the flux change as you move further away from a light source?
　　　What is the central concept of this sub-section?
**Absolute Visual Magnitude**
　　　Define absolute visual magnitude.
　　　Why is the absolute visual magnitude useful?
　　　What is the central concept of this sub-section?
**Luminosity**
　　　What is the luminosity of a star?
　　　Do all stars have the same luminosity?
　　　What is the central concept of this sub-section?

What are the fundamental ideas presented in this section?

**8-3    The Diameters of Stars**
**Luminosity, Radius, and Temperature**
　　　What two factors affect a star's luminosity?
　　　What property of a star is plotted along the vertical axis of a Hertzsprung - Russell diagram?
　　　What property of a star is plotted along the horizontal axis of a Hertzsprung - Russell diagram?
　　　Where in the H-R diagram are cool stars plotted?
　　　What is the name of the region in the H-R diagram where 90% of the stars are plotted?
　　　Where in the H-R diagram are the giants, supergiants, red dwarfs, and white dwarfs plotted?
　　　What is the central concept of this sub-section?
**Luminosity Classification**
　　　List the roman numeral designation for each of the luminosity classes and give the name of each class.
　　　What is the central concept of this sub-section?
**Spectroscopic Parallax**
　　　What is spectroscopic parallax?
　　　What part of a star's spectrum tells us the luminosity of the star?
　　　What is the central concept of this sub-section?

What are the fundamental ideas presented in this section?

**8-4    The Masses of Stars**
**Binary Stars in General**
Which star in a binary system is closest to the center of mass?
What two properties of a binary system's orbit do we need to know to determine the mass of the binary system?
What factors make it difficult to determine the mass of a binary system?
What is the central concept of this sub-section?
**Window on Science 8-1.  Learning About Nature Through Chains of Inference**
How do scientists use chains of inference?
**Visual Binary Systems**
What are visual binary systems?
Are the periods of visual binary systems usually long or short?
What is the central concept of this sub-section?
**Spectroscopic Binary Systems**
What do we use to analyze the motions in a spectroscopic binary?
Can the two stars of a spectroscopic binary system be seen separately from Earth?
Look at Figure 8-14 on page 146, at what times are the two stars moving with the same radial velocities?
What is the central concept of this sub-section?
**Eclipsing Binary Systems**
What is an eclipsing binary system?
Can the two stars in an eclipsing binary system be seen separately?
What is plotted in a light curve?
In the light curve of an eclipsing binary star, what does the depth of the eclipse depend upon?
What is the central concept of this sub-section?

What are the fundamental ideas presented in this section?

**8-5    A Survey of the Stars**
**Mass, Luminosity, and Density**
Where on the main sequence in the H-R diagram are the least massive stars plotted?
What is the mass-luminosity relationship?
For which luminosity class(es) does the mass-luminosity relationship apply?
What three groups do stars fall into when we calculate their densities?
What is the central concept of this sub-section?
**Window on Science 8-2.  Information About Nature: Basic Scientific Data**
According to this Window on Science, why do scientists spend so much time collecting data?
**Surveying the Stars and The Family of Stars on pages 152 and 153**
The text states that if we are not careful when we survey the stars, we will get "biased results".
What is meant by biased-results?
What two problems do we face in surveying the stars near the sun?
What is the most common type of star near the sun?
What is the central concept of this sub-section?

What are the fundamental ideas presented in this section?

## KEY CONCEPTS

This chapter describes how we know what we know about the fundamental properties of stars and what we can learn from these properties. Stars have six fundamental properties, mass, luminosity, radius, temperature, composition and age. Each of these is unique to the star and clearly defines the star. This chapter describes how luminosity, radius, and mass are determined and how they can be used to draw some conclusions about stars in general.

The luminosity of a star tells us how much energy the star produces each second, and so tells us about what is going on inside the star. Determining the luminosity requires determining the distance to the star. The first section discusses the method used to determine the distance to nearby stars. The method of parallax is important because it shows that it is possible to measure the distances to stars, and it also is the foundation upon which all other astronomical distances are based, such as spectroscopic parallax.

The relationship between luminosity, temperature and radius helps us understand that stars of similar temperatures can have very different radii. This leads to the luminosity classification, which will be used in the following chapters.

The major concept of the fourth section is the methods for determining the mass of a star. It is only from binary stars that we can accurately determine the mass of individual stars. Eclipsing binaries are the most useful in that the masses and radii of the individual stars in the binary can be determined from the light curve and radial velocity data. The visual binaries allow for an accurate determination of the mass of each of the stars in the system. The spectroscopic binaries exist in abundance, but permit only a lower limit on the mass of the stars to be determined.

The mass-luminosity relation is discussed in the fifth section, and this relation should be emphasized. It will be used in the discussions of stellar evolution and in determining the age of stellar clusters. This final section also looks at how we can determine the relative abundances of stars of various spectral types and luminosity classes. It is difficult to accurately survey the stars because the most luminous stars are very rare, while the most common stars are very faint. In order to do an accurate survey we must include all stars within a given distance of the sun. If we go far enough out to include a sampling of the O and B main sequence stars, then we can no longer detect the spectral type M main sequence stars. Therefore, it is difficult to get an unbiased survey.

The importance of the H-R diagram must be clearly understood. It will play a prominent role in the chapters that follow. It is vital that you understand that the H-R diagram is a tool that allows us to quickly see relations between stars' temperatures, luminosities, and diameters.


## WEBSITES OF INTEREST

http://instruct1.cit.cornell.edu/courses/astro101/java/binary/binary.htm    Spectroscopic Binary demo
http://zebu.uoregon.edu/~js/ast122/lectures/lec05.html    Stellar Properties Tutorial
http://www.physics.sfasu.edu/astro/binstar.html    Binary Star demonstrations


## QUESTIONS ON CONCEPTS

1.  Given three stars of the same spectral type, how can their luminosities be determined?

    Solution:
    The luminosity class of a star can be determined from the width of certain spectral lines. Notice in Figure 8-9 on page 141 that the width of the spectral lines in the supergiant are very narrow, while the lines in the giant are wider, and the lines in the main-sequence star are extremely broad.

2.  What properties of a star can be determined if the star is part of an eclipsing binary system and how can these properties be determined?

Solution:
In an eclipsing binary system we can determine the mass, temperature, radius, and luminosity of each star.  It is only for stars in eclipsing binary systems that we can determine all of this information in a relatively straightforward manner.  The temperature of each star can be determined from the spectrum of the system.  When neither star is eclipsed the spectrum shows the red shifted spectrum of one of the stars, and the blue shifted spectrum of the other.  Analysis of the spectrum then tells us the spectral type and temperature of each star.  We can also determine the composition of each star once the temperature is determined.  The mass can be determined from the light curve and radial velocity variations.  The radial velocity variations tell us the orbital speed, and the radial velocity variations and light curve both tell us the orbital period.  From this we can determine the true separation of the two stars and use Kepler's law to determine the mass of the system.  The radius of the stars can be determined from the light curve.  Notice in the figure below that the time it takes to go from point 2 to point 3 is related to the diameter of star A.  Since we know the orbital velocity of each star and the time it takes to go from one point to the next, we can find the diameter of Star A.  Additionally, the diameter of Star B can be found from measuring the time it takes to go from point 2 to point 4.  In this case, Star A has moved a distance equal to the diameter of Star B, so we can calculate Star B's diameter.

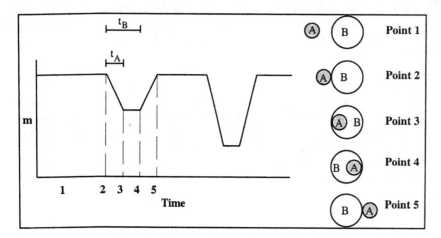

Finally, the luminosity of the two stars can be determined in two ways.  First we can determine the luminosity from the spectrum of each star.  The widths of the lines tell us the luminosity class.  Additionally, if we know each star's radius and temperature, we can calculate the star's radius.  From this it is easy to see that eclipsing binary systems are extremely valuable to astronomers.

3. Use the H-R diagram below to answer the following questions.
   a.   Which stars have the same temperature?
   b.   Which stars have the same luminosity?
   c.   Which star has the largest radius?
   d.   Which star is most like the sun?

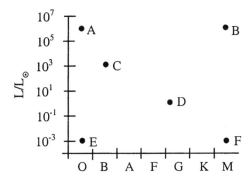

Solution:
The H-R diagram plots the luminosity and temperature of stars. The luminosity increases as you move vertically upward in the diagram. The temperature increases as you move horizontally to the left. So high luminosity stars are found at the top of the diagram, hot stars at the left side, and cool stars at the right.

Solution to part a:
Stars with the same temperature will be along the same vertical line and have the same spectral type. Stars A and E have the same temperature and are the hottest stars plotted. Stars B and F have equal temperatures, but are much cooler than Stars A and E.

Solution to part b:
Stars with the same luminosity will be located on the same horizontal line. Stars A and B have identical luminosities and are the two most luminous stars plotted. Stars E and F have identical luminosities, but are much less luminous than all other stars plotted.

Solution to part c:
The luminosity of a star depends on its radius and its temperature. Consider two stars with the same luminosity but different temperatures, like Star A and Star B. Both stars put out the same amount of energy, but the hot one, Star A, puts out more energy from each square meter of its surface. Therefore, if the cool star, Star B, is to put out as much energy as a hotter star, it must have more square meters of surface, that is, it must be larger in radius. From this we can see that Star B is larger than Star A and Star F is larger than Star E.

Now consider two stars of the same temperature but different luminosities, like Star A and Star E. Both stars radiate the same amount of energy from each square meter of surface, so the one with the larger luminosity, Star A, will have the larger radius. From this we see that Star A is larger in radius than Star E and Star B is larger in radius than Star F.

Pulling this all together, Star E is smaller than both Star A and Star F. Star B is larger than both Star A and Star F, which are both larger than Star E.

So the star with the largest radius is Star B, and the star with the smallest radius is Star E. From the information given it isn't clear whether Star F or Star A is larger. Generally, radius increases toward the upper right in the H-R diagram.

Solution to part d:

The star most like the sun will have a temperature and luminosity like the sun. Note that if its temperature and luminosity are the same as the sun's, then its radius must also be equivalent. Additionally, since the sun is on the main sequence, any object with the same temperature and luminosity as the sun is also on the main sequence and will have the same mass as well.

The sun is a G2 star with a luminosity of 1 $L_{sun}$. The only G star plotted is Star D and its luminosity is very close to 1 $L_{sun}$. Star D is most like the sun.


## WORKED EXAMPLES OF PROBLEMS REQUIRING MATHEMATICS

1. What is the distance to a star with a measured parallax of 0.05 seconds of arc?

   Solution:
   We know the parallax angle = $p$ = 0.05 seconds of arc.
   The distance in parsecs, $d(pc)$, can be found when the parallax angle is known by using the parallax formula in *By the Numbers 8-1* on page 135.

   $$d(pc) = \frac{1}{p('')} = \frac{1}{0.05''} = 20 \text{ parsecs}$$

   The distance to the star is 20 parsecs.

2. What is the parallax angle of a star at a distance of 15 parsecs?

   Solution:
   We know the distance to the star = $d$ = 15 parsecs.
   The parallax angle, $p$, can be calculated when the distance is know by using the parallax formula in *By the Numbers 8-1* on page 135,

   $$d(pc) = \frac{1}{p('')}$$

   We want to solve for $p$, the parallax angle, so we will multiply both sides by $p$ and then divide by the distance, $d$.

   $$p('') \cdot d(pc) = p('') \cdot \frac{1}{p('')} = 1. \text{ Now divide by the distance, } d.$$

   $$\frac{p('') \cdot d(pc)}{d(pc)} = \frac{1}{d(pc)}$$

   $$p('') = \frac{1}{d(pc)} = \frac{1}{15 \text{ pc}} = 0.067 \text{ seconds of arc}$$

   The parallax angle is 0.067 seconds of arc for a star at a distance of 15 pc.

3. A star has an apparent visual magnitude of 6.8 and an absolute visual magnitude of –3.2. What is the distance to this star?

   Solution:
   We know:
   apparent visual magnitude = $m_v$ = 6.8, and
   absolute visual magnitude = $M_v$ = –3.2.
   We can use the distance modulus found in *By the Numbers 8-2* on page 137 to find the distance to this star.

   $$d(\text{pc}) = 10^{(m_v - M_v + 5)/5} = 10^{(6.8 - (-3.2) + 5)/5} = 10^{(6.8 + 3.2 + 5)/5} = 10^{(15)/5} = 10^3 = 1,000$$

   The distance to the star is 1000 pc.

4. A star is located at a distance of 2,500 pc and has an absolute visual magnitude of 2.6. What is the apparent visual magnitude of this star?

   Solution:
   We know:
   the distance to the star = $d$ = 2,500 pc, and
   the absolute visual magnitude = $M_v$ = 2.6.
   The distance can be found by using the distance modulus in *By the Numbers 8-2* on page 137.

   $$m_v - M_v = -5 + 5\log_{10}(d(\text{pc}))$$

   $$m_v = M_v - 5 + 5\log_{10}(d(\text{pc})) = 2.6 - 5 + 5\log_{10}(2500\text{pc})$$

   The $\log_{10}$ function on most calculators is a key labeled $\boxed{\text{LOG}}$. To find the $\log_{10}(2500)$, enter 2500 on your calculator and then push the $\boxed{\text{LOG}}$ key. The display should read $\boxed{3.39794}$. Therefore,

   $$m_v = 2.6 - 5 + 5\log_{10}(2500\text{pc}) = -2.4 + 5 \cdot (3.40) = -2.4 + 17.0 = 14.6$$

   The apparent visual magnitude of the star is 14.6.

5. A star has a radius of 10 solar radii and a temperature of 4,000 K.
   a. What is the ratio of the luminosity of this star to that of the sun?
   b. The sun's luminosity is $3.8 \times 10^{26}$ J/s, what is the luminosity of this star?

   Solution to part a:
   We know:
   Radius of the star = $R$ = 10 $R_{\text{sun}}$,
   Temperature of the star = $T$ = 4,000 K,
   Radius of the sun = 1 $R_{\text{sun}}$, and
   Temperature of the sun = $T_{\text{sun}}$ = 5,800 K.
   We can find the luminosity of the star using the luminosity, radius, and temperature relationship in *By the Numbers 8-3* on page 139.

   $$\frac{L}{L_{\text{sun}}} = \left(\frac{R}{R_{\text{sun}}}\right)^2 \cdot \left(\frac{T}{T_{\text{sun}}}\right)^4 = \left(\frac{10\,R_{\text{sun}}}{1\,R_{\text{sun}}}\right)^2 \cdot \left(\frac{4000}{5800}\right)^4 = (100) \cdot (0.6897)^4 = (100) \cdot (0.226) = 22.6$$

   The luminosity of the star is 22.6 solar luminosities.

Solution to part b:
We know:
Luminosity of the star = $L = 22.6 \, L_{sun}$ and
Luminosity of the sun = $L_{sun} = 3.8 \times 10^{26}$ J/s.
If we substitute the value for the luminosity of the sun into the first equation we find the luminosity of the star in J/s.

$$L = 22.6 \, L_{sun} = (22.6) \cdot \left(3.8 \times 10^{26} \text{ J/s}\right) = 8.6 \times 10^{27} \text{ J/s}.$$

The luminosity of the star is equal to 22.6 solar luminosities or $8.6 \times 10^{27}$ J/s.

6. A binary system has a period of 40 years and an average separation of 50 AU. What is the mass of the binary system?

Solution:
We know:
Period of the orbit = $P$ = 40 years and
Average separation distance = $a$ = 50 AU.
We can use Newton's laws of motion and gravity in *By the Numbers 8-3* on page 143 to solve for the mass of the system, $M_A + M_B$.

$$M_A + M_B = \frac{a^3}{P^2} = \frac{(50 \text{ AU})^3}{(40 \text{ years})^2} = \frac{125,000}{1600} = 78 \text{ solar masses}$$

The mass of the binary system is 78 solar masses.

7. What is the luminosity of an eight solar mass main sequence star?

Solution:
We know the mass of the main sequence star = $M$ = 8 solar masses.
We can use the mass-luminosity relation in *By the Numbers 8-4* on page 151 to find the luminosity of a main sequence star.

$$L = M^{3.5} = 8^{3.5} = 8 \cdot 8 \cdot 8 \cdot \sqrt{8} = 512 \cdot 2.83 = 1450 \text{ solar luminosities}$$

The luminosity of an eight solar mass main sequence star is approximately 1,450 solar luminosities, which is $5.5 \times 10^{29}$ J/s.


**MORE MATH RELATED PROBLEMS**
(Answers at the end of the chapter)

1. What is the distance to a star with a parallax of 0.062 seconds of arc?

2. What is the parallax angle of a star at a distance of 120 parsecs?

3. Complete the following table.

| Star | Parallax angle (seconds of arc) | Distance (parsecs) |
|---|---|---|
| Capella | 0.072 | |
| β Crucis | | 150.3 |
| Rigel Kentaurus | | 1.32 |
| Arcturus | 0.091 | |
| Achenar | 0.30 | |
| Aldebaran | | 20.9 |

4. What is the distance to a star with an apparent magnitude of 11.2 and an absolute magnitude of 15.3?

5. What is the absolute visual magnitude of a star at a distance of 25 pc with an apparent visual magnitude of 3.6?

6. Complete the following table using the equations in *By the Numbers 5-1* and *By the Numbers 5-2*.

| Star | $m_V$ | $M_V$ | d (pc) | Parallax (sec of arc) | Spectral Type |
|---|---|---|---|---|---|
| Proxima Cen | 11.05 | 15.45 | | | M5 |
| α Cyg | | −7.1 | 491 | | A2 |
| Sirius | −1.47 | | 2.67 | | A1 |
| ε Ind | 4.7 | | | 0.29 | K5 |
| Capella | | −0.6 | | 0.072 | G8 |

7. What is the luminosity of a star with a radius 1000 times smaller than the sun's, and temperature of 20,000 K?

8. What is the mass of a binary system if it has an orbital period of 5 years and an average separation distance of 10 AU?

9. What is the luminosity of a main sequence star with a mass of 0.1 solar masses?

10. What is the mass of a main sequence star with a luminosity of 2,200 solar luminosities?


**PRACTICE TEST**
**Multiple Choice Questions**

1. The parallax of a star can be measured if
   a. the star's apparent magnitude can be measured.
   b. we can obtain a good spectrum of the star.
   c. the star is an eclipsing binary.
   d. the star is on the main sequence.
   e. the star is within about 100 pc of the sun.

2. _____ can be used to determine the distance to a star when the spectrum of the star can be used to determine its spectral type and luminosity class.
   a. Spectroscopic parallax
   b. The distance modulus
   c. The mass luminosity relation
   d. Hertzsprung-Russell diagram
   e. Visual binaries

3. At what point in the light curve to the right is the hot star in front of the cool star?
   a. 1
   b. 2
   c. 3
   d. 4
   e. 5

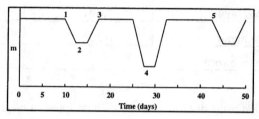

4. If two stars have the same luminosity and Star A is twice as far away as Star B, which of the following is true?
   a. Star A will appear brighter than Star B.
   b. Star A will have a larger absolute visual magnitude than Star B.
   c. Star A will have a smaller absolute visual magnitude than Star B.
   d. Star A will appear 4 times fainter than Star B.
   e. Star A and Star B will have the same apparent visual magnitude.

5. The stars on the main sequence all have about the same
   a. mass.
   b. density.
   c. luminosity.
   d. apparent magnitude.
   e. temperature.

6. It is difficult to accurately determine the relative number of stars of different spectral types and luminosity classes because
   a. the stars are only visible during the night.
   b. the stars are all about the same absolute visual magnitude.
   c. there is a large variation in the luminosities of stars.
   d. the temperatures of the stars are too similar.
   e. we can not determine the luminosity of nearby stars.

7. What is the distance to a star with a parallax of 0.05 seconds of arc?
   a. 0.05 parsecs
   b. 50 parsecs
   c. 500 parsecs
   d. 20 parsecs
   e. 50 light years

8. Spica is a B1 V star. Based on this information which of the following are true?
   I.   Spica has a surface temperature greater than the sun.
   II.  Spica has a mass that is greater than that of the sun.
   III. Spica is more luminous than the sun.
   IV.  Spica is located near the upper left hand corner in the H-R diagram.

   a. I & II
   b. II & IV
   c. II, III, & IV
   d. I, II, & III
   e. I, II, III, & IV

9. The binary system has an orbital period of 10 years and an average separation distance of 5 AU. What is the total mass of the system?
   a. 1.25 solar masses
   b. 40 solar masses
   c. 4 solar masses.
   d. 0.8 solar masses.
   e. 15 solar masses.

10. What is the distance to a star with an apparent visual magnitude of 3.6 and an absolute visual magnitude of 1.6?
    a. 0.25 pc
    b. 2.5 pc
    c. 25 pc
    d. 25 ly
    e. 2.5 ly

**Fill in the Blank Questions**

11. A 2 solar mass star on the main sequence would have a luminosity of approximately _____ solar luminosities.

12. A spectral type M3 V star is _____ massive than the sun.

13. The hydrogen lines in spectral type A stars are the most narrow for luminosity class _____ stars.

14. Luminosity class V objects are known as _____.

15. On the H-R diagram below, indicate the location of a red supergiant.

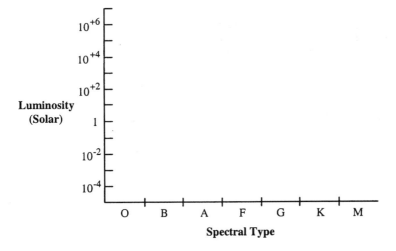

## True-False Questions

16. For stars on the main sequence, the luminosity can be estimated by the formula $L = M^{3.5}$.

17. The absolute visual magnitude and the apparent visual magnitude of a star are equal if the star is at a distance of 10 pc.

18. From the location of a star in an H-R diagram one can estimate the star's radius.

19. The total mass of a binary system can be calculated from the average separation distance between the two stars and the distance to the binary.

20. A spectroscopic binary shows periodic variations in its radial velocity.

## ADDITIONAL READING

Boss, Alan P., "The Birth of Binary Stars." *Sky and Telescope 97* (June 1999), p. 32.

Evans, D. S., T. G. Barnes, and C. H. Lacy "Measuring Diameters of Stars." *Sky and Telescope 58* (Aug. 1979), p. 130.

Henry, Todd J. "Brown Dwarfs Revealed - At Last!" *Sky and Telescope 91* (April 1996), p. 24.

Trefil, James "Putting Stars in their Place." *Astronomy 28* (Nov. 2000), p. 62.

## ANSWERS TO MORE MATH RELATED PROBLEMS
1. 16.1 pc
2. 0.0083 seconds of arc
3. Answers appear in **bold**.

| Star | Parallax angle (seconds of arc) | Distance (parsecs) |
|---|---|---|
| Capella | 0.072 | **13.9** |
| β Crucis | **0.00665** | 150.3 |
| Rigel Kentaurus | **0.758** | 1.32 |
| Arcturus | 0.091 | **11.0** |
| Achenar | 0.30 | **3.3** |
| Aldebaran | **0.0478** | 20.9 |

4. 1.5 pc
5. 1.6
6. Complete the following table using the equations in *By the Numbers 5-1* and *By the Numbers 5-2*.

| Star | $m_V$ | $M_V$ | d (pc) | Parallax (sec of arc) | Spectral Type |
|---|---|---|---|---|---|
| Proxima Cen | 11.05 | 15.45 | **1.32** | **0.76** | M5 |
| α Cyg | **1.36** | −7.1 | 491 | **0.0020** | A2 |
| Sirius | −1.47 | **1.40** | 2.67 | **0.37** | A1 |
| ε Ind | 4.7 | **7.01** | 3.45 | 0.29 | K5 |
| Capella | **0.11** | −0.6 | **13.9** | 0.072 | G8 |

7. $0.00014\ L_{sun} = 5.4\times10^{22}$ J/s
8. 40 solar masses
9. $3.2\times10^{-4}\ L_{sun} = 1.2\times10^{23}$ J/s
10. 9.0 solar masses

**ANSWERS TO PRACTICE TEST**

1.  e
2.  a
3.  b
4.  d
5.  b
6.  c
7.  d
8.  e
9.  a
10. c
11. 11
12. less
13. I, or supergiants
14. dwarfs or main sequence stars
15. Spectral type M and luminosity of $10^5$ to $10^6$ solar luminosities.
16. T
17. T
18. T
19. F
20. T

# CHAPTER 9

# THE FORMATION AND STRUCTURE OF STARS

## UNDERSTANDING THE CONCEPTS

**9-1    The Birth of Stars**
   **The Interstellar Medium and Three Kinds of Nebulae on pages 160 and 161**
      What is an HII region?
      What causes a reflection nebula?
      What are dark nebula?
      What evidence supports the existence of an interstellar medium?
      In what ways does interstellar affect light from distant stars?
      What is the central concept of this sub-section?
   **The Formation of Stars from the Interstellar Medium**
      What force works to collapse a cloud of gas?
      What keeps most clouds from collapsing?
      What can cause a cloud of gas to begin to collapse?
      What four mechanisms can produce the shockwaves that might initiate the collapse of a gas cloud?
      What is the central concept of this sub-section?
   **The Formation of Protostars**
      What is a protostar?
      Throughout contraction, what is the source of a protostar's thermal energy?
      What determines how long it takes a protostar to collapse to form a main-sequence star?
      What is the central concept of this sub-section?
   **Window on Science 9-1.  Understanding Science: Separating Facts from Theories**
      According to this Window on Science, what is the fundamental work of science?
      What is the difference between a scientific fact and a scientific theory?
      Can one theory be tested against another theory?
      What is the central concept of this sub-section?
   **Observations of Star Formation and Observational Evidence of Star Formation on pages 166 and 167**
      Why is it difficult to locate and observe protostars?
      What evidence suggests that protostars form surrounded by disks of gas and dust?
      What is a T Tauri Star?
      What do bipolar flows tell us about star formation?
      What is the central concept of this sub-section?

   What are the fundamental ideas presented in this section?

**9-2    Fusion in Stars**
   **The CNO cycle**
      What is the CNO cycle?
      For which stars is the CNO cycle important?
      What temperature is required for the CNO cycle to operate?
      What is the central concept of this sub-section?

### Heavy-Element Fusion
Which fusion process occurs after hydrogen fusion in most stars?
What is the triple-alpha process?
What temperatures are required for helium and carbon fusion?
What is the central concept of this sub-section?

### The Pressure-Temperature Thermostat
What is meant by the pressure-temperature thermostat?
Explain how a star is able to produce just the right amount of energy to balance the gravitational force.
The stability of a star depends on the balance between what two forces?
What is the central concept of this sub-section?

What are the fundamental ideas presented in this section?

## 9-3    Stellar Structure
### The Laws of Mass and Energy
What is the law of conservation of mass?
What is the law of conservation of energy?
What is the central concept of this sub-section?

### Hydrostatic Equilibrium
What is hydrostatic equilibrium?
The pressure in a gas depends on two things, what are they?
What other mechanism is hydrostatic equilibrium related to?
What is the central concept of this sub-section?

### Energy Transport
Name three mechanisms that transport thermal energy?
What are the two most important energy transport mechanisms inside a star?
What is convection?
What is radiation?
What is opacity?
Why is convection important in stars?
What is the central concept of this sub-section?

### Stellar Models
What are the four laws of stellar structure?
How are stellar models constructed and what form do they take?
What is the central concept of this sub-section?

### Window on Science 9-2.  Quantitative Thinking with Mathematical Models
What is one of the most powerful tools in science?
What is a mathematical model?
What limits the reliability of a mathematical model?
What is the central concept of this sub-section?

What are the fundamental ideas presented in this section?

## 9-4    Main Sequence Stars
### The Mass-Luminosity Relation
Why is there a mass-luminosity relationship?
What are brown dwarfs?
What is the central concept of this sub-section?

### The Life of a Main Sequence Star
What happens in the core of a main sequence star as it converts hydrogen into helium?
Throughout its life on the main-sequence, how will a star's luminosity and temperature change?
What factor determines the amount of time a star spends on the main-sequence?
What is the central concept of this sub-section?

What are the fundamental ideas presented in this section?

## 9-5    The Orion Nebula
### Evidence of Young Stars and Star Formation in the Orion Nebula on pages 178 and 179

> Why are we able to see the Orion nebula when most of the region is composed of a dark molecular cloud?
>
> What unique feature do most of the stars in the Orion nebula have that stars like the sun do not possess?
>
> Why does there need to be a very hot star in order to produce the emission nebula seen in Orion?
>
> What does the existence of a disk around a star in a nebula tell us about the age of the star?
>
> What is the distance across the Orion nebula?

What are the fundamental ideas presented in this section?


## KEY CONCEPTS

This chapter describes the interstellar medium, how stars form from that material, and how stars on the main sequence produce energy and control the production of energy.

The first-half of the first section presents the evidence for the existence of material between the stars, that is the interstellar medium. The primary observations discussed to support the existence of the interstellar medium include the observations of emission, reflection, and dark nebulae, interstellar reddening, and interstellar absorption lines. Emission and dark nebulae show us where stars are born, but also make it beautifully clear that there are large complexes of gas and dust in amongst the stars. Interstellar reddening is a direct result of small solid grains that scatter blue light much better than they scatter red light. Consequently, light that passes through a long column of these dust grains will have much of the blue light removed and they will appear redder than if the dust grains were not there. The interstellar absorption lines tell us that there is gas distributed throughout interstellar space. This is a very low-density gas.

The second half of the first section describes our understanding of how stars form from the gas and dust of the interstellar medium. Stars originate in dark molecular clouds, which begin to condense when they encounter a shock wave. As fragments within the cloud condense their cores increase in temperature and pressure until they begin to produce energy by nuclear fusion. There is a great deal of evidence to support these ideas. Our observations of T Tauri stars, Bok globules, Herbig-Haro objects and bipolar flows show us objects in various stages of star formation, from the collapsing fragments of a Bok globule to the T Tauri stars about to initiate nuclear fusion.

The second section talks about nuclear fusion in stars. Understanding the thermonuclear fusion of hydrogen to helium is critical to understanding stellar evolution because it controls the longest portion of a star's life. Additionally, other nuclear reactions will occur in most stars later in their lives that are controlled in the same way that the hydrogen reaction is controlled.

Main-sequence stars produce their energy in one of two hydrogen fusion processes. For stars about the mass of the sun and smaller, energy is produced primarily by the proton-proton chain, which is described in chapter seven. For stars more massive than the sun, they produce most of their energy while on the main sequence by the CNO cycle. Both the CNO cycle and the proton-proton chain produce energy by converting four hydrogen nuclei into one helium nucleus. Other nuclear fuels are available to stars, but all are used after the stars leave the main sequence.

One of the most important concepts presented in this chapter is the pressure-temperature thermostat. The outward flow of energy produces the pressure needed to balance the inward pull of gravity. The amount of energy produced in a nuclear reaction depends strongly on the temperature of the gas. If the temperature increases, so does the outward flow of energy and the pressure. This pressure will cause the star to expand. For most gases the pressure, temperature and density of the gas are related in a very simple manner. If the star expands, it will cool and as it cools the amount of energy produced will decrease, so the pressure will drop and the star will shrink. So the link between the amount of energy produced and the temperature and density of the gas, and the link between the pressure, temperature and density of the gas, conspire to control the rate at which energy flows outward to exactly balance the inward pull of the gravitational force.

The third section describes the processes we use to construct mathematical models of the interiors of stars. One of those processes, hydrostatic equilibrium, must be understood. It is the process that helps us understand all of the changes that occur in a star as it evolves from a dark molecular cloud to the main sequence and ultimately to its final state.

The final section of this chapter is a nice discussion of how understanding a single object requires observations at many wavelengths. The Orion nebula is a beautiful object that can be easily viewed with binoculars even from large cities. Taking the time to understand this one star formation region provides you with an opportunity to see how all of the concepts of this chapter can be used to understand a very beautiful and complicated celestial object.


## WEBSITES OF INTEREST

http://hubblesite.org/newscenter/                            Hubble Space Telescope.
http://www.aao.gov.au/                                        Anglo-Australian Observatory.


## QUESTIONS ON CONCEPTS

1.  Why are distant stars redder than nearby stars of the same spectral type?

    Solution:
    The spectral type of a star tells us the star's temperature. The spectral type is based on the absorption lines that are present in the star's spectrum and not on the continuous part of the star's spectrum. The strengths of these lines are not affected by dust that lies between the star and the observer. The color of the star is associated with the continuous part of the star's spectrum, and dust can affect this continuous part of the spectrum. Small grains of dust in the interstellar medium preferentially scatter blue light out of the beam of light coming to the observer. This means that the blue end of the continuous spectrum will be less intense than expected because the dust is scattering some of this blue light. Therefore the star will have a redder color than is expected from the star's spectral type.

    The amount of blue light scattered by the dust depends on how many dust grains the light must travel through. If the star is a long ways away, its light will have to travel through a larger amount of the dust than the light from a closer star. In this way the distant star will look redder than a nearby star of the same spectral type and temperature.

2.  According to the stellar model in Figure 9-13, at what fraction of the sun's mass and radius is 100% of the sun's luminosity produced?

    Solution:
    The model is found in the table in Figure 9-13 on page 173. The luminosity is recorded in the fifth column of the table. The luminosity is given in terms of the total luminosity of the sun, so the location where $L/L_\odot$ is equal to 1.00 tells us the outer edge of the region where the luminosity is generated inside the sun. Looking at the table we find that the luminosity reaches 1.00 $L_\odot$ at a radius of 0.40 $R_\odot$ and at a mass of 0.82 $M_\odot$. So all of the luminosity of the sun is produced in the inner 40% of the sun's radius and this radius contains 89% of the sun's mass.

## WORKED EXAMPLES OF PROBLEMS REQUIRING MATHEMATICS

1. What is the main sequence lifetime of a 30 solar mass star?

   Solution:
   We know the mass of the star = $M$ = 30 solar masses.
   We can use the *By the Numbers 9-1* on page 176 to find the lifetime of the star on the main sequence.
   Let $T$ = the lifetime of the star on the main sequence.

   $$T = \frac{1}{M^{2.5}} \cdot (\text{the lifetime of the sun}) = \frac{1}{30^{2.5}} \cdot \left(1.0 \times 10^{10} \text{ years}\right) = \frac{1}{30 \cdot 30 \cdot \sqrt{30}} \cdot \left(1.0 \times 10^{10} \text{ years}\right)$$

   $$T = \frac{1}{4930} \cdot \left(1.0 \times 10^{10} \text{ years}\right) = \frac{\left(1.0 \times 10^{10} \text{ years}\right)}{4930} = 2.0 \times 10^{6} \text{ years}$$

   A 30 solar mass star will spend $2.0 \times 10^{6}$ years (only 2 million years) on the main sequence.

2. A typical brown dwarf has a temperature of 2,500 K, at what wavelength does it radiate the most energy?

   Solution:
   We know the temperature of the object = $T$ = 2,500 K.
   We can use Wien's law from *By the Numbers 6-1* page 98 to find $\lambda_{max}$, the wavelength at which the maximum energy is radiated.

   $$\lambda_{max} = \frac{3,000,000 \text{ nm} \cdot \text{K}}{T} = \frac{3,000,000 \text{ nm} \cdot \text{K}}{2,500 \text{ K}} = 1,200 \text{ nm}$$

   The wavelength at which the most energy is radiated is 1,200 nm.

3. The interstellar medium dims starlight by 1.9 magnitudes for each 1000 pc it travels through. What fraction of a stars light reaches Earth for a star at a distance of 2,630 parsecs?

   Solution:
   We know three things:
   the distance to the star = 2,630 pc,
   we lose 1.9 magnitudes for each 1000 pc, and
   the intensity decreases by a factor of 2.512 for each 1 magnitude lost.

   Since the star is 2,630 pc away and light is lost at the rate of 1.9 magnitudes per 1000 pc, we find that we will lose

   $$magnitudes\ lost = \frac{1.9 \text{ mag}}{1000 \text{ pc}} \cdot 2,630 \text{ pc} = 1.9 \text{ mag.} \cdot \frac{2,630 \text{ pc}}{1,000 \text{ pc}} = 1.9 \text{ mag.} \cdot 2.63 = 5.0 \text{ mag.}$$

   In *By the Numbers 2-1* on page 13 we see that a difference of 5 magnitudes means that the intensity ratio decreases by 100. We can also use the formula

   $$\frac{I_A}{I_B} = 2.512^{(m_B - m_A)} = 2.512^{5} = 100,$$

where I have chosen $I_A$ to be the intensity of the light we would have received, and $I_B$ to be the intensity of the light we did receive. So the fraction of the total that we receive is the inverse of this. <u>The fraction of light we receive is then 0.01 of what we would if the dust were not present.</u>

4.  On an image of the constellation of Orion, the Orion nebula appears to have an angular diameter of about 3,600 seconds of arc, or 1 degree. The Orion nebula is at a distance of approximately 400 pc. Based on this information, what is the linear diameter of the Orion nebula?

    <u>Solution:</u>
    We know
    the distance to the Orion nebula = 400 pc and
    the angular diameter of the nebula = 3,600 seconds of arc = 3,600".
    We will use the small angle formula in *By the Numbers 3-1* on page 32.

    $$\frac{angular\ diameter}{206,265"} = \frac{linear\ diameter}{distance} \quad or \quad \frac{linear\ diameter}{distance} = \frac{angular\ diameter}{206,265"}$$

    If we multiply both sides by the distance, we can solve for the linear diameter.

    $$distance \cdot \frac{linear\ diameter}{distance} = distance \cdot \frac{angular\ diameter}{206,265"}$$

    $$linear\ diameter = distance \cdot \frac{angular\ diameter}{206,265"} = 400\,pc \cdot \frac{3,600"}{206,265"} = 400\,pc \cdot 0.0175 = 7.0\,pc$$

    <u>The nebula is about 7.0 parsecs in diameter.</u>

## MORE MATH RELATED PROBLEMS
(Answers at the end of the chapter)

1.  What is the main-sequence lifetime of a 0.5 solar mass star?

2.  How long will a 40 solar mass star burn hydrogen to helium in its core?

3.  If dust in the interstellar medium decreases the light we receive by 1.9 magnitudes per 1000 pc, what fraction of the light would we receive from a star at 100? (This one is difficult.)

4.  If dust in the interstellar medium decreases the light we receive by 1.9 magnitudes per 1000 pc, what fraction of the light would we receive from a star at 8,500 pc (the distance to the center of our galaxy)? (This one is difficult.)

5.  IC 430 is an emission nebula at a distance of 49 parsecs. It has an angular diameter of 660 arc seconds, what is the linear diameter of IC 430?

6.  The Cocoon nebula has an angular diameter of 12 arc minutes and is 1700 parsecs away. What is the linear diameter of the Cocoon nebula?

7.  NGC 7748 is an emission nebula in the constellation of Cetus, the sea monster. It has an angular diameter of 3 arc minutes and is 330 parsecs away. What is the linear diameter of NGC 7748?

1. Ninety percent of all stars fuse _____ and lie on the main sequence in the H-R diagram.
   a. helium to form carbon
   b. hydrogen to form helium
   c. carbon and nitrogen to form oxygen
   d. helium to form hydrogen
   e. carbon to form nitrogen and oxygen.

2. Stars with masses greater than the sun
   a. use the CNO cycle to produce energy while on the main sequence.
   b. have longer main sequence lifetimes than the sun.
   c. have smaller luminosities than the sun while on the main sequence.
   d. do not show interstellar reddening.
   e. have temperatures less than the sun's while on the main sequence.

3. _____ occurs when the pressure from the outward flow of energy in a layer is equal to the force due to gravity pushing down on that layer.
   a. Conservation of mass
   b. Opacity
   c. Hydrostatic equilibrium
   d. The CNO cycle
   e. Conservation of energy

4. Why are massive stars on the main sequence more luminous than low mass stars on the main sequence?
   a. The massive star are less dense than the low mass stars.
   b. The massive stars are primarily hydrogen, while the low mass stars are primarily helium.
   c. The massive stars are primarily helium, while the low mass stars are primarily hydrogen.
   d. The massive stars are spinning faster which increases the outward flow of energy.
   e. The massive stars must create a greater outward flow of energy to support their larger mass.

5. A shock wave from a supernova can cause
   a. emission nebulae to become a dark nebula.
   b. low mass stars to burn their nuclear fuels more slowly.
   c. high mass stars to burn their fuels more slowly.
   d. molecular clouds to fragment and begin to collapse.
   e. bipolar flows in main-sequence stars.

6. Why are interstellar absorption lines so narrow?
   a. The gas in the interstellar medium is very hot.
   b. The gas in the interstellar medium contains a lot of dust.
   c. The gas in the interstellar medium is predominantly molecules.
   d. The gas in the interstellar medium is very low density.
   e. The gas in the interstellar medium comes from supergiants, which have very narrow spectral lines.

7. A mathematical model of the sun tells us that
   a. most of the energy from the sun is transported to the surface by conduction.
   b. most of the mass of the sun is located near its surface.
   c. most of the energy is produced with in the central 25% of the sun's radius.
   d. the density of the sun changes very little from the core to the photosphere.
   e. none of the above

8. Bok globules are
   a. small, dark fragments of the interstellar medium.
   b. only found near low mass stars.
   c. produce the energy needed to make an emission nebula produce its light.
   d. components of a reflection nebula.
   e. never found near regions of star formation.

9. The dust in the interstellar medium
   a. is composed of grains much larger than the wavelengths of visible light.
   b. causes interstellar reddening.
   c. produces the interstellar absorption lines.
   d. does not affect the brightness of distant stars.
   e. all of the above

10. What is the approximate main-sequence lifetime of a 7 solar mass star?
    a. 7 solar lifetimes
    b. 130 solar lifetimes
    c. 910 solar lifetimes
    d. 0.14 solar lifetimes
    e. 0.0077 solar lifetimes

## Fill in the Blank Questions

11. For a(n) _____ _____ to exist there must be a young hot star ($T \geq 25,000$ K) relatively nearby.

12. Energy transport by _____ is important in many stars because it mixes the material in the star, as well as transporting energy.

13. Small grains of dust in reflection nebula preferentially scatter _____ photons.

14. Objects with masses less than about _____ solar masses cannot initiate hydrogen fusion.

15. The _____ of a gas depends on the temperature and density of the gas.

## True-False Questions

16. Absorption lines due to interstellar gas are narrower than the lines from stars because the gas in the interstellar medium has a lower pressure than the gas in stars.

17. Conservation of mass states that the total luminosity of the star is equal to the sum of the energy generated in each of the shells.

18. Opacity is a measure of the resistance of a gas to the flow of radiation.

19. The main sequence has a limit at the lower end because there is a minimum temperature for hydrogen fusion.

20. There is a mass-luminosity relation because all stars on the main sequence produce energy by the CNO cycle.

## ADDITIONAL READING

Davidson, Kris "Crisis at Eta Carina?" *Sky and Telescope 95* (Jan. 1998), p. 36.

Greenberg, J. Mayo "The Secrets of Stardust" *Scientific American 283* (Dec. 2000), p. 70.

Kaler, James B. "Eyewitness to Stellar Evolution." *Sky and Telescope 97* (March 1999), p. 40.

Nesme-Ribes, Elizabeth, Sallie L. Baliunas, and Dmitry Sokoloff "The Stellar Dynamo." *Scientific American 277* (Aug. 1996), p. 46.

Ray, Thomas P. "Fountains of Youth: Early Days in the Life of a Star." *Scientific American 283* (Aug. 2000), p. 42.

## ANSWERS TO MORE MATH RELATED PROBLEMS
1. 5.7 solar lifetimes = $5.7 \times 10^{10}$ years
2. $9.9 \times 10^{-5}$ solar lifetimes = $9.9 \times 10^5$ years
3. $1/1.191 = 0.84$
4. $1/(2,900,000) = 3.5 \times 10^{-7}$
5. 0.16 pc
6. 5.9 pc
7. 0.29 pc

## ANSWERS TO PRACTICE TEST
1. b
2. a
3. c
4. e
5. d
6. d
7. c
8. a
9. b
10. e
11. emission nebula or HII region
12. convection
13. blue or short wavelength
14. 0.08
15. pressure
16. T
17. F
18. T
19. T
20. F

# THE DEATHS OF STARS

## UNDERSTANDING THE CONCEPTS

### 10-1   Giant Stars
#### Expansion into a Giant
What happens to a star when it is no longer able to produce energy by the fusion of hydrogen in its core?

What causes the outer layers of a star like the sun to expand after it quits fusing hydrogen in its core?

What happens to the surface temperature of stars as they evolve away from the main sequence?

What is the central concept of this sub-section?

#### Degenerate Matter
Under normal circumstances in a gas, what does the pressure of the gas depend upon?

What is the Pauli exclusion principle?

What is degenerate matter?

In a degenerate gas, pressure no longer depends on temperature, why is this important?

What is the central concept of this sub-section?

#### Helium Fusion
What is the minimum temperature needed initiate helium fusion?

For what stellar masses do the electrons in the core become degenerate before helium fusion begins?

What is helium flash?

What nuclei does helium fusion produce?

What is the central concept of this sub-section?

#### Window on Science 10-1.  Causes in Nature: The Very Small and the Very Large
Why do astronomers study very small things like atoms, nuclei, and electrons?

What is the central concept of this sub-section?

#### Star Clusters: Evidence of Evolution and Star Cluster H-R Diagrams on pages 188 and 189
What are the two different types of star clusters and how do they differ?

How can astronomers determine the age of a star cluster?

Why is the location of the main sequence for a globular cluster different than that of "normal" stars?

What is the central concept of this sub-section?

What are the fundamental ideas presented in this section?

### 10-2   The Deaths of Low- and Medium-Mass Stars
#### Red Dwarfs
What role does convection play in the evolution of red dwarfs?

Why can't a red dwarf become a giant?

How long are red dwarfs expected to remain on the main sequence?

What is the central concept of this sub-section?

#### Medium-Mass Stars
What is the range in mass of the medium-mass stars?

What nuclear fuels do medium mass stars use?

What is the central concept of this sub-section?

**Planetary Nebulae and The Formation of Planetary Nebulae on pages 192 and 193**
What is the range in mass for stars that form planetary nebulae?
What is the approximate lifetime for a planetary nebula?
Describe a simple model for how planetary nebulae are formed.
What processes have been suggested to explain some of the asymmetries seen in planetary nebulae?
What is the central concept of this sub-section?

**White Dwarfs**
What are typical temperatures, radii, and densities for a white dwarf?
What keeps a white dwarf from collapsing?
What is the composition of the interior of a typical white dwarf?
What is the future of a white dwarf?
What is the Chandrasekhar limit?
What is the central concept of this sub-section?

What are the fundamental ideas presented in this section?

**10-3    The Evolution of Binary Systems**
**Mass Transfer**
What are Roche lobes?
What is the inner Lagrangian point?
When can mass be transferred rapidly from one star to another in a binary system?
What is the central concept of this sub-section?

**Evolution with Mass Transfer**
What is the Algol paradox?
What is the central concept of this sub-section?

**Accretion Disks**
What is angular momentum?
Why do accretion disks form?
What two things happen to the gas in an accretion disk?
What is the central concept of this sub-section?

**Novae**
What causes a nova?
Why are novae expected to repeat?
What is the central concept of this sub-section?

**The End of the Earth**
What is the fate of the sun?
Why might the sun not produce a planetary nebula?
What is the central concept of this sub-section?

What are the fundamental ideas presented in this section?

**10-4    The Deaths of Massive Stars**
**Nuclear Fusion in Massive Stars**
Which nuclear fuels will the massive stars use to produce energy?
Why does fusion of the heavy elements happen much faster than the fusion of hydrogen (two reasons)?
What is the central concept of this sub-section?

**Supernova Explosions of Massive Stars**
Why can't massive stars produce energy from the fusion of iron?
What mechanism expands the out layers of the star during a supernova?
What is the central concept of this sub-section?

**Types of Supernovae**
What are the differences between Type I and Type II supernovae?
What are the differences between Ia and Ib supernovae?
What is the central concept of this sub-section?

### Observations of Supernovae
What is the Crab nebula?

Why is SN1987A important?

What type of star exploded to form SN1987A?

How is a supernova remnant formed?

How is synchrotron radiation produced?

What is the central concept of this sub-section?

### Window on Science 10-2. Astronomical Images, False Color, and Reality
What are some reasons that false color images are used in astronomy?

What is the central concept of this sub-section?

What are the fundamental ideas presented in this section?

## KEY CONCEPTS

Chapter 10 answers the question: "What happens to a star after it runs out of hydrogen in its core?" The most important aspect of this chapter is that the evolution of a star depends critically on its mass and only on its mass. The mass of a star determines the extent to which it can squeeze the core and generate certain core temperatures, which determines the nuclear reactions that occur.

Medium-mass stars become giants, but on their way they develop degenerate cores. That is the density of the core becomes so great that the electrons become degenerate. The degenerate electrons cannot easily change their energies; consequently the pressure of the electron gas does not depend on the temperature. The degeneracy of the electrons in the cores of these stars leads to helium flash. Helium flash can be understood only if you clearly understand the pressure-temperature thermostat and the fact that the rate of nuclear fusion increases as the temperature increases. In degenerate material the pressure of the gas does not depend on temperature so the pressure-temperature thermostat cannot regulate the nuclear reaction until the degenerate state of the matter is reversed. Helium flash is not an observable phenomenon because all of the energy that is produced in the core is absorbed in the massive envelope of the star. Finally, intermediate mass stars will produce white dwarfs, which are again controlled by the presence of degenerate electrons.

The third section of this chapter discusses how binary stars evolve. It is from binary star evolution that novae and Type I supernovae are produced. Additionally, over half of the stars in the universe are believed to be in multiple star systems, so binary star evolution is more common than single star evolution.

Massive stars avoid the consequences of degenerate matter by heating their cores more rapidly as they collapse so that helium begins to fuse before the electrons become degenerate. In these stars it is important to understand how supernovae occur. As a massive star begins to fuse iron in its core, a shock wave develops near the core and moves outward through the envelope and rips the outer layers of the star apart. A supernova is a violent explosion and many of our theories on how they are produced have been confirmed by observations of Supernova 1987A.

## WEBSITES OF INTEREST

| | |
|---|---|
| http://rsd-www.nrl.navy.mil/7212/montes/sne.html | Extensive links on supernovae. |
| http://zebu.uoregon.edu/galaxy.html | Astro. Dept., U of Oregon |
| http://www.aao.gov.au/images.html | Anglo-Australian Observatory. |

**QUESTIONS ON CONCEPTS**

1.  What can be determined about the age of the cluster in the H-R diagram below?

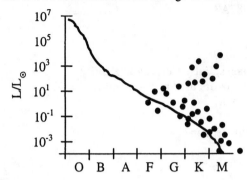

Solution:

The age of a star cluster can be determined from the location of the main sequence turnoff point. The turnoff point in this cluster is located at a luminosity of about 10 solar luminosities. This means that main sequence stars that had luminosities greater than 10 solar luminosities have used up their hydrogen and left the main sequence. Those stars with luminosities less than 10 solar luminosities are still burning hydrogen in their cores.

So we can find the age of the cluster if we can determine the main sequence lifetime of a star that has a main sequence luminosity of 10 solar luminosities. We know that this star will be more massive than the sun because it has a greater luminosity and is on the main sequence like the sun. Since it has a greater mass, it will have a lifetime less than that of the sun. Therefore, the star at the turnoff point has an age less than $1.0 \times 10^{10}$ years. So this cluster is less than 10 billion years old.

Calculating the exact mass and lifetime of a main sequence star with a luminosity of 10 solar luminosities is a little beyond the scope of most general astronomy courses, but it is possible with the mass-luminosity relationship and the equation for the lifetime of a main sequence star. Doing it completely gets a little confusing, so let's take a guess and check it. The mass-luminosity relation tells us that the luminosity is equal to the mass to 3.5 power. So if we guess a mass of 10 solar masses, we find that the luminosity is over 3000 solar luminosities, which is much to big. So the mass is between 1 and 10 solar masses, and probably closer to 1 than 10. Let's try 2 (it helps to know the answer ahead of time). 2 raised to the 3.5 power is 11.3. This means that the luminosity of a 2 solar mass star is 11.3 solar luminosities, and 11.3 is close enough to 10 solar luminosities for our estimate. This tells us that the turnoff point is at a mass of about 2 solar masses. We can find the age of cluster using the lifetime equation of *By the Numbers 9-1*. This is worked out below in the first problem in the "Worked Examples of Problems requiring Mathematics". We find that the age of the cluster is approximately $1.8 \times 10^9$ years, that is, 1.8 billion years.

2.  Why is degenerate matter important to astronomers?

Solution:

The central reason that degenerate matter is important to astronomers is that it supplies a tremendous force that opposes both contraction and expansion. It can support a great deal of weight, and it can restrain a great deal of pressure. This is possible because degenerate electrons cannot change their energies easily. They can neither slow down to decrease their energy nor speed up to increase their energy, because most of the possible energy states are full. Only when a great deal of energy is added can they jump reverse this degenerate state. In the case of medium mass stars expanding to become giants, the electrons in the core become degenerate. When this happens, the core will continue to contract and increase in temperature as new material is added to it. Eventually the temperature will become hot enough to begin helium fusion, which releases large amount of energy. The core cannot expand and cool, because the electrons cannot change their energies. Consequently the temperature of the core increases rapidly and produces more and more energy. Eventually sufficient energy is absorbed by the electrons to reverse the degenerate state. The core will then cool and regulate the nuclear reaction by the pressure-temperature thermostat.

In white dwarfs the degenerate electrons cannot be easily compressed and are capable of supporting the mass of the star. There is a limit to the amount of mass that the degenerate electrons can support and this limits the mass of a white dwarf to 1.4 solar masses, the Chandrasekhar limit.

3. What causes a main sequence star to expand to become a red giant?

Solution:
Any expansion in a star is caused by the outward flow of energy. As a star exhausts hydrogen in its core, hydrogen fusion begins in a shell around the core. The core is contracting because it no longer produces energy by fusion. Without the outward flow of energy from fusion, the core contracts and increases in temperature. The shell of hydrogen increases in temperature and produces a large amount of energy. This energy is not balanced by gravity and the outer layers of the star expand and cool. Therefore the radius of the star increases, and the surface temperature decreases.

## WORKED EXAMPLES OF PROBLEMS REQUIRING MATHEMATICS

1. If the stars at the turnoff point in a cluster have a mass of 2.0 solar masses, what is the age of the cluster?

Solution:
The age of a star cluster is equal to the lifetime of the main sequence stars at the turnoff point in the cluster.
We know the mass of the main sequence stars at the turnoff point = $M = 2.0$ solar masses.

We can use *By the Numbers 9-1* on page 176 to calculate the lifetime of a two solar mass star, which will be equal to the age of the cluster.

$$T = \frac{1}{M^{2.5}} \cdot \left(1.0 \times 10^{10} \text{ years}\right) = \frac{1}{2.0^{2.5}} \cdot \left(1.0 \times 10^{10} \text{ years}\right) = \frac{1}{5.66} \cdot \left(1.0 \times 10^{10} \text{ years}\right) = 1.8 \times 10^{10} \text{ years}$$

The age of the cluster with a turnoff point at 2.0 solar masses is $1.8 \times 10^9$ years.

2. IC2149 is a planetary nebula located about 870 pc from Earth. It has an angular diameter of 15 seconds of arc. What is the linear diameter of the nebula?

Solution:
We know
angular diameter of the nebula = 15" and
the distance to the nebula = 870 pc.

We can use the small angle formula from *By the Numbers 3-1* on page 32 to find the linear diameter.

$$\frac{\textit{angular diameter}}{206,265"} = \frac{\textit{linear diameter}}{\textit{distance}} \quad \text{or} \quad \frac{\textit{linear diameter}}{\textit{distance}} = \frac{\textit{angular diameter}}{206,265"}$$

If we multiply both sides by the distance, we can solve for the linear diameter.

$$distance \cdot \frac{linear\ diameter}{distance} = distance \cdot \frac{angular\ diameter}{206,265"}$$

$$linear\ diameter = distance \cdot \frac{angular\ diameter}{206,265"} = 870\,pc \cdot \frac{15"}{206,265"} = 870\,pc \cdot 0.000073 = 0.063\,pc$$

The nebula is about 0.063 parsecs in diameter.

3. If the gas in IC2149 is expanding at a rate of 18 km/s, how old is the nebula? Use the linear diameter from the problem above.

Solution:
We know:
the diameter of the nebula = 0.063 pc,
the rate at which the nebula is expanding = 18 km/s, and
the number of kilometers in one parsec, that is 1 pc = $3.1 \times 10^{13}$ km.

What we want to find is the time it has taken the nebula to travel a distance equal to the **radius** of the nebula, which is half of 0.063 pc or 0.0315 pc. It expands at the rate of 18 km/s. To calculate the time it takes we must first convert the distance traveled to kilometers.

$$d\,(km) = d\,(pc) \cdot (number\ of\ kilometers\ in\ 1\,pc) = 0.0315\,pc \cdot 3.15 \times 10^{13}\,\tfrac{km}{pc} = 9.8 \times 10^{11}\,km$$

So the gas must travel a distance of $9.8 \times 10^{11}$ km at a rate of 18 km/s. The time it takes is equal to the distance traveled divided by the speed.

$$time\,(s) = \frac{distance\,(km)}{speed\,(km/s)} = \frac{9.8 \times 10^{11}\,km}{18\,\tfrac{km}{s}} = 5.4 \times 10^{10}\,s$$

This can be converted to years by dividing by the number of seconds in a year.

$$time\,(years) = \frac{time\,(s)}{(\#\ of\ seconds/minute) \cdot (\#\ of\ minutes/hour) \cdot (\#\ of\ hours/day) \cdot (\#\ of\ days/year)}$$

$$time\,(years) = \frac{5.4 \times 10^{10}\,s}{60\,\tfrac{s}{min} \cdot 60\,\tfrac{min}{hr} \cdot 24\,\tfrac{hr}{day} \cdot 365.25\,\tfrac{days}{year}} = \frac{5.4 \times 10^{10}\,s}{3.16 \times 10^{7}\,\tfrac{s}{year}} = 1700\ years$$

The nebula has taken 1700 years to reach its present size.

4. If a star swells to a supergiant and becomes 400 times larger in diameter, by what factor does the star's average density change? (Hint: The volume of a sphere is equal to $\frac{4}{3}\pi r^3$.)

Solution:
We know that the mass of the star does not change and that the final radius is equal to 400 times the initial radius. This tells us that the density is going to decrease as the star expands, so the final density must be less than the initial density.

If we let $r_i$ equal the initial radius, and $r_f$ equal the final radius, the $r_f = 400r_i$. We also know that the density is the mass divided by the volume.

$$d = \frac{mass}{volume} = \frac{M}{\frac{4}{3}\pi r^3}$$

We want to find the ratio of the two densities. Let $density_i$ equal the initial density, and $density_f$ equal the final density. Then

$$density_f = \frac{M}{\frac{4}{3}\pi r_f^3} \quad \text{and} \quad density_i = \frac{M}{\frac{4}{3}\pi r_i^3}$$

$$\frac{density_f}{density_i} = \frac{\left(\frac{M}{\frac{4}{3}\pi r_f^3}\right)}{\left(\frac{M}{\frac{4}{3}\pi r_i^3}\right)} = \left(\frac{M}{\frac{4}{3}\pi r_f^3}\right) \cdot \left(\frac{\frac{4}{3}\pi r_i^3}{M}\right) = \frac{M\frac{4}{3}\pi r_i^3}{M\frac{4}{3}\pi r_f^3} = \frac{r_i^3}{r_f^3} \quad \text{so} \quad \frac{density_f}{density_i} = \frac{r_i^3}{r_f^3}$$

This is a general result that is applicable to any problem of this type. We can now insert the values for this specific problem.

$$\frac{density_f}{density_i} = \frac{r_i^3}{r_f^3} = \frac{r_i^3}{(400 r_i)^3} = \frac{r_i^3}{400^3 \cdot r_i^3} = \frac{1}{400^3} = \frac{1}{6.4 \times 10^7}$$

The final density is $6.4 \times 10^7$ times smaller than the initial density.

## MORE MATH RELATED PROBLEMS
(Answers at the end of the chapter)

1.  The Pleiades is an open cluster with a turnoff point at about 6.3 solar masses. What is the approximate age of the Pleiades?

2.  A cluster has a turnoff point at a mass of 0.9 solar masses, what is the approximate age of this cluster?

3.  NGC1501 is a planetary nebula with an angular diameter of about 50 arc seconds and is located at a distance of 4170 pc. The nebulae is expanding at a rate of 35 km/s.
    a.  What is the linear diameter of the nebula?
    b.  What is the age of the nebula?

4.  NGC7354 is a planetary nebula located in Cepheus. It is 111 pc away, has an angular diameter of 32 seconds of arc, and is expanding at a rate of 60 km/s.
    a.  What is the linear diameter of the nebula?
    b.  What is the age of the nebula

5.  If a star is compressed from its main sequence radius to a radius 100 times smaller, by what factor has the star's density changed?

6.  A star expands to form a giant 60 times its original radius, by what factor has its density changed?

**PRACTICE TEST**
**Multiple Choice Questions**

1.  Mass can flow from one star in a binary system to its companion through the first _____ point.
    a.  Pauli
    b.  Lagrangian
    c.  Hubble
    d.  Chandasekhar
    e.  Seeds

2.  The Chandrasekhar limit exists because
    a.  helium requires a temperature of at least 100 million K to initiate fusion.
    b.  angular momentum keeps material from falling directly on to a white dwarf.
    c.  stars with masses less than 0.08 cannot squeeze their cores enough to initiate hydrogen fusion.
    d.  iron is the most tightly bound of all nuclei.
    e.  degenerate electrons cannot support more than about 1.4 solar masses of material.

3.  A star cluster has a turnoff point at a spectral type of K5.  What do you know about this star cluster?
    a.  The cluster is older than 10 billion years.
    b.  The cluster is younger than 10 billion years.
    c.  The cluster won't contain any white dwarfs.
    d.  The cluster won't contain any red dwarfs.
    e.  none of the above

4.  As a white dwarf cools its radius will not change because
    a.  pressure due to nuclear reactions in a shell just below the surface keeps it from collapsing.
    b.  pressure does not depend on temperature because the white dwarf is too hot.
    c.  pressure does not depend on temperature because the star has exhausted all its nuclear fuels.
    d.  pressure does not depend on temperature for a white dwarf because the electrons are degenerate.
    e.  material accreting onto it from a companion maintains a constant radius.

5.  Why can't a massive star generate energy through iron fusion?
    a.  Massive stars cannot squeeze their core hard enough for iron nuclei to combine.
    b.  Massive stars do not possess any iron to fuse.
    c.  Massive stars cores are convective and don't allow iron nuclei to collide.
    d.  Iron is the most tightly bound nuclei and requires more energy to combine than fusion produces.
    e.  Iron fusion is an important source of energy in massive stars.

6.  In the diagram to the right, which point indicates the location on the H-R diagram of a one solar mass star when it is composed of carbon and oxygen nuclei and degenerate electrons?
    a.  1
    b.  2
    c.  3
    d.  4
    e.  5

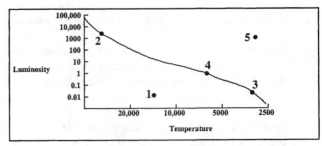

7.  Which of the following nuclear fuels does a 0.2 solar mass star use over the course of its entire evolution?
    a.  only hydrogen
    b.  hydrogen and helium
    c.  hydrogen, helium and carbon
    d.  hydrogen, helium, carbon, and neon
    e.  hydrogen, helium, carbon, neon, and oxygen.

8. _____ occurs when material on the surface of a white dwarf initiates hydrogen fusion.
   a. An accretion disk
   b. Helium flash
   c. The triple-alpha process
   d. A nova
   e. The Algol paradox

9. Convection in the interior of a red dwarf
   a. allows degenerate electrons to accumulate in the core of the star.
   b. allows degenerate electrons to accumulate in the outer layers of the star.
   c. allows helium fusion to begin without going through helium flash.
   d. keeps the electrons in the core from becoming degenerate.
   e. keeps the material mixed and doesn't allow a core of helium ash to form.

10. If the stars at the turnoff point of a cluster have a mass of 6 solar masses, what is the age of the cluster?
    a. $6.0 \times 10^{10}$ years
    b. $1.7 \times 10^{9}$ years
    c. $1.1 \times 10^{8}$ years
    d. $8.8 \times 10^{11}$ years
    e. The age of a star cluster cannot be determined from the mass of stars at the turnoff point.

**Fill in the Blank Questions**

11. _____ _____ occurs as medium mass star become giants because degenerate electrons in the core do not allow the core to expand as it heats up.

12. _____ radiation is produced by high-velocity electrons moving through a magnetic field.

13. The _____ exclusion principle states that no two particles in a gas can have exactly the same energy.

14. No known white dwarf has a mass greater than the _____ limit.

15. The central star in a planetary nebula will become a _____ _____.

**True-False Questions**

16. A supernova is the expulsion of the outer layers of a moderate mass star that has a degenerate carbon and oxygen core.

17. A star will experience a helium flash if its core is degenerate when helium ignites.

18. In degenerate matter the pressure depends only on the temperature.

19. The lowest mass object that can initiate thermonuclear fusion of hydrogen has a mass of about 0.08 solar masses.

20. No neutrinos were detected from supernova 1987A, confirming our theory that white dwarfs cannot be pushed over the Chandrasekhar limit by adding mass to them.

## ADDITIONAL READING

Chaboyer, Brian C. "Rip Van Twinkle" *Scientific American 284* (May 2001), p. 44.

Davidson, Kris "Crisis at Eta Carina?" *Sky and Telescope 95* (Jan. 1998), p. 36.

Frank, Adam "Bursting the Bubbles." *Astronomy 28* (April 2000), p. 38.

Irion, Robert "Pursuing the Most Extreme Stars." *Astronomy 27* (Jan. 1999), p. 48

Kahabka, Peter, Edward P. J. van den Heuvel, and Saul Rappaport "Supersoft X-ray Stars and Supernovae." *Scientific American 280* (Feb. 1999), p. 46.

Kaler, James B. "Eyewitness to Stellar Evolution." *Sky and Telescope 97* (March 1999), p. 40.

Kirshner, Robert P. "Supernova 1987A: The First Ten Years." *Sky and Telescope 93* (Feb. 1997), p. 35.

Zimmerman, Robert "Scoping Out the Monster" *Astronomy 28* (Feb. 2000), p. 38.

## ANSWERS TO MORE MATH RELATED PROBLEMS

1.  0.01 solar lifetimes = $1.0 \times 10^8$ years
2.  1.3 solar lifetimes = $1.3 \times 10^{10}$ years = 13 billion years
3.  a.   1.0 pc
    b.   $9.0 \times 10^{11}$ s = 28,000 years
4.  a.   0.017 pc
    b.   $8.9 \times 10^9$ s = 280 years
5.  density will be $1 \times 10^6$ times greater after collapse.
6.  density will be $2.2 \times 10^5$ times smaller after expansion.

## ANSWERS TO PRACTICE TEST

1.  b
2.  e
3.  a
4.  d
5.  d
6.  a
7.  a
8.  d
9.  e
10. c
11. Helium flash
12. Synchrotron
13. Pauli
14. Chandrasekhar
15. white dwarf
16. F
17. T
18. F
19. T
20. F

# C H A P T E R   1 1

# NEUTRON STARS AND BLACK HOLES

## UNDERSTANDING THE CONCEPTS

### 11-1  Neutron Stars

**Theoretical Prediction of Neutron Stars**

What is a neutron star?

What happens inside the collapsing core of a massive star if the core's mass is greater than 1.4 solar masses?

What is the maximum mass of a neutron star?

How do the temperature, density, rotation rate, and magnetic field of neutron stars compare to the same quantities of the sun?

What is the central concept of this sub-section?

**The Discovery of Pulsars**

What type of telescope was used to discover the first pulsars?

Why were white dwarfs ruled out as producers of the pulses observed by Bell and Hewish?

What is the central concept of this sub-section?

**A Model Pulsar and The Lighthouse Model of a Pulsar on pages 212 and 213**

Why is the rapid rotation of a neutron star necessary in the light house model?

Why is a strong magnetic field necessary in the light house model?

How might a neutron star operate like the lighthouse model and we not see it as a pulsar?

What is the central concept of this sub-section?

**The Evolution of Pulsars**

Why do pulsars slow down?

What mechanisms might produce a supernova remnant that does not contain a neutron star?

What is the central concept of this sub-section?

**Binary Pulsars**

How did the binary pulsar PSR1913+16 indicate that gravitational radiation exists?

How does Hercules X-1 produce X-rays?

What are X-ray bursters?

What is the central concept of this sub-section?

**The Fastest Pulsars**

How can some old pulsars be spinning very rapidly if pulsars slow down with time?

What is the central concept of this sub-section?

**Window on Science 11-1.  The Impossibility of Proof in Science**

Why do scientists claim that proof of a theory is impossible?

What is the central concept of this sub-section?

**Pulsar Planets**

What is the suggested origin of the planets that orbit pulsar PSR1257+12?

What is the central concept of this sub-section?

What are the fundamental ideas presented in this section?

## 11-2 Black Holes

### Escape Velocity
Upon what two things does the escape velocity of a star or planet depend?
What is the central concept of this sub-section?

### Schwarzschild Black Holes
What is the event horizon?
What is the Schwarzschild radius of a black hole?
Would Earth be sucked in if the sun became a black hole?
What is the central concept of this sub-section?

### A Leap into a Black Hole
What is time dilation?
What is a gravitational red shift?
Why do things falling into a black hole get very hot?
What is the central concept of this sub-section?

### The Search for Black Holes
What types of objects do we use in our search for black holes?
Does the evidence at this point support the existence of black holes?
What is the central concept of this sub-section?

### Window on Science 11-2. Natural Checks on Fraud in Science
Why is fraud in science rare?
Why is peer-review important in publishing scientific results?
What is the central concept of this sub-section?

### Jets of Energy from Compact Objects
How can compact objects, including black holes radiate large amounts of energy?
What processes can cause compact objects to eject high speed jets of matter?
What is the central concept of this sub-section?

### Gamma-Ray Bursts
What is a gamma-ray burster?
What are two theories currently being considered as mechanisms for producing gamma-ray bursters?
What is the central concept of this sub-section?

What are the fundamental ideas presented in this section?


## KEY CONCEPTS

This chapter centers on two of the most fascinating types of objects in the universe, neutron stars and black holes. The key concept of section one is that collapsing cores of stars with masses in excess of 1.4 solar masses cannot balance gravity with the pressure due to electron degeneracy. Such objects will collapse further to form neutron stars. Neutron stars are very low luminosity objects and difficult to observe unless they also happen to be pulsars. The lighthouse model of a pulsar helps us see what might be going on to produce the pulses observed from pulsars. However, it is a model with many of the details yet to be worked out. Consequently, our understanding of how pulsars create the beams of radiation is still very uncertain. Pulsars have become very important in astronomy and physics because binary pulsars have allowed us to test the predictions of the general theory of relativity and to locate the first extra-solar planets.

Black holes are one type of object that most students want to know more about, and many students have misconceived ideas about these objects. The key concept of the section is how black holes can be detected and this means understanding how material accretes onto a black hole. Traveling into or through a black hole is something that is often discussed as if it is possible to survive the trip, which it is not. However, putting yourself in the gravitational field of a black hole, mentally, can help you understand what happens to matter as it encounters a black hole. The mental exercise of placing yourself in the midst of some phenomenon is often employed by astronomers and physicists to try to understand what is happening. This exercise helps us see that matter will build up around a black hole to form a large and very hot accretion disk. This disk should be hot enough to radiate large amounts of X-rays. So the search for black holes relies on finding X-rays. Additionally we need to be able determine that the object is not some other compact object like a neutron star or white dwarf. We can do that if we can determine the mass of the object. White dwarfs must be less than 1.4 solar masses and neutron stars must be less than about 3 solar masses. So if we can find a binary system that has an unseen companion with a mass in excess of 3 solar masses that emits copious amounts of X-rays, we will have a black hole candidate. Several such objects have been found, and most astronomers are confident that black holes do exist with masses of 5 to 10 solar masses. Black holes will be encountered again in chapters 12 and 14, so understanding the ways in which material accreting onto black holes produces detectable energy is important.

Students often get the impression that neutron stars have masses greater than 1.4 solar masses and black holes must have masses greater than 3 solar masses. This is not true. The Chandrasekhar limit states that the maximum mass of a white dwarf is 1.4 solar masses. This sets a limit on the maximum mass of a white dwarf, but does not set a minimum on the mass of a neutron star or black hole. If we want to find a black hole we need to find a compact object with a mass greater than 3 solar masses so we can be sure it isn't a neutron star, but this does not mean that black holes can't have masses less than 3 solar masses.

## WEBSITES OF INTEREST

http://cosmology.berkeley.edu/Education/BHfaq.html
http://antwrp.gsfc.nasa.gov/htmltest/rjn_bht.html

Answers to questions on black holes.
Virtual trips to black holes and neutron stars.

## QUESTIONS ON CONCEPTS

1.  Why do we not expect to find a 5 solar mass neutron star?

    Solution:
    Neutron stars support their mass by the pressure of the degenerate neutrons. Because most of the energy levels available to the neutron are filled, the neutrons cannot slow down and collapse further. So as more mass is added, they can support it up to a point. The limit of the force that degenerate neutrons can support is not well established, but we do know that it is around the equivalent of 2 or 3 solar masses. Therefore it is unlikely that any neutron star could have a mass greater than 3 solar masses.

2.  What types of objects do we use in our search for black holes?

    Solution:
    The search for black holes requires that we be able to determine the mass of the back hole so that we can be sure that it isn't a neutron star or white dwarf. We can only determine the mass of a star if it is in a binary system, so we need to look at binary systems. It will be easier to estimate the mass of the compact object if the visible star in the binary system is a low to medium mass main sequence star. Since we can't see both stars in the binary system we can only determine the total mass of the system. If the visible star is a low to medium mass main sequence star we can estimate its mass fairly accurately, this is not true for giants, supergiants, or high mass main sequence stars. Additionally we know that when material falls toward a compact object like a neutron star or black hole, the matter will pile up and form an accretion disk that will be very hot and produce large amounts of X-rays. Therefore we want to observe X-ray binary systems, and in particular such systems that contain a visible low to medium mass main sequence star.

## WORKED EXAMPLES OF PROBLEMS REQUIRING MATHEMATICS

1.  A neutron star rotates 30 times per second and has a diameter of 10 km, what is the velocity of a particle on its equator? (Hint: the circumference of the star is equal to $\pi d$.)

    Solution:
    We know:
    the rotation period $= t = 1/30 = 0.033$ seconds and
    the diameter of the star $= d = 10$ km.
    The speed of an object is the distance it travels divided by the time it takes to travel that distance. An object on the equator of the neutron star will travel a distance equal to the star's circumference in one rotation.

    $$velocity = \frac{distance\ traveled}{time} = \frac{circumference}{period} = \frac{\pi d}{t} = \frac{\pi \cdot 10\,\text{km}}{0.033\,\text{s}} = 952\,\text{km/s}$$

    The velocity of a particle on the surface of the neutron star is 950 km/s.

2.  What is the Schwarzschild radius of a 6 solar mass black hole?

    Solution:
    We know the mass of the black hole $= M = 6$ solar masses.
    We can use the formula for the Schwarzschild radius on page 220, but we must first convert the mass into kilograms. From Table A-5 in the appendix, we find that 1 solar mass $= 2.0 \times 10^{30}$ kg.

    $$M\,(\text{kg}) = M\,(\text{solar masses}) \cdot (\text{kilograms per solar mass}) = (6\,\text{solar masses}) \cdot \left(2.0 \times 10^{30}\,\tfrac{\text{kg}}{\text{solar mass}}\right) = 1.2 \times 10^{31}\,\text{kg}$$

    Now we can use the formula for the Schwarzschild radius.

    $$R_s = \frac{2GM}{c^2} = \frac{2 \cdot \left(6.67 \times 10^{-11}\,\frac{m^3}{s^2 kg}\right) \cdot \left(1.2 \times 10^{31}\,kg\right)}{\left(3.0 \times 10^8\,\frac{m}{s}\right)^2} = 18{,}000\,\text{m} = 18\,\text{km}$$

    The Schwarzschild radius of a 6 solar mass black hole is 18 km.

    Look at Table 11-1 on page 221 and the solution here. Notice that the Schwarzschild radius in km is always equal to 3 times the mass of the star in solar mass units. That is

    $$R_S\,(\text{km}) = 3 \cdot M\,(\text{solar masses})$$

    This relationship can be used to simplify calculations of the Schwarzschild radius, which is also the radius of the event horizon.

3.  The binary pulsar PSR1913+16 has a period of 7.75 hours. If the average separation distance is about 0.01 AU, what is the total mass of the system?

    Solution:
    We know
    The average separation distance $= r = 0.01$ AU, and
    the period of the orbit $= P = 7.75$ hours.
    We can use Newton's laws of gravity and motion from *By the Numbers 8-4* on page 143 to solve for the total mass. In order to use this formula, we must convert the period into years. We know that there are 24 hours/day and 365.25 days/year so

$$P_{\text{years}} = \frac{7.75 \text{ hours}}{24 \frac{\text{hours}}{\text{day}} \cdot 365.25 \frac{\text{days}}{\text{year}}} = 0.000884 \text{ years}$$

Now we use the formula from *By the Numbers 8-4* to find the mass.

$$M_A + M_B = \frac{r_{\text{AU}}^3}{P_{\text{years}}^2} = \frac{(0.01 \text{ AU})^2}{(0.000884 \text{ years})^2} = 1.3 \text{ solar masses}$$

So the total mass of the system is 1.3 solar masses.

4. An X-ray binary consists of a G2 V star and a compact X-ray source. The system has a period of 10 days and the average separation distance between the star and X-ray source is 0.2 AU.
   a. What is the total mass of the system?
   b. Is the compact companion a good candidate to be a black hole?

Solution to part a:
We know
the period of the binary system = $P$ = 10 days = 0.0274 years and
the average separation distance = $r$ = 0.2 AU.
We can use Newton's laws of gravity and motion from *By the Numbers 8-4* on page 143 to solve for the total mass.

$$M_A + M_B = \frac{r_{\text{AU}}^3}{P_{\text{years}}^2} = \frac{(0.2 \text{ AU})^3}{(0.0274 \text{ years})^2} = 10.7 \text{ solar masses}$$

The mass of the binary system is 10.7 solar masses.

Solution to part b:
We know
the mass of the system = 10.7 solar masses and
the visible star is a G2 V.
Our sun is a G2 V star, which means it is a spectral type G2 main sequence star. So the visible star has a mass equal to the mass of the sun. The mass of the unseen companion must then be equal to $10.7 - 1.0 = 9.7$ solar masses. Since the upper limit for the mass of a neutron star is about 3 solar masses, the compact companion is an excellent black hole candidate.


**MORE MATH RELATED PROBLEMS**
(Answers at the end of the chapter)

1. What is the speed of an object on the equator of a white dwarf if the white dwarf has a radius of 6,000 km and a rotation period of 1 hour.

2. What is the speed of a particle on the sun's equator? The sun rotates once every 30 days and has a radius of $7.0 \times 10^5$ km.

3. What is the Schwarzschild radius of a 1 million solar mass black hole?

4. How far from the center of an eight solar mass black hole is the event horizon?

5. What is the mass of the binary system that has an orbital period of 12 hours and a separation distance of 0.02 AU?

6. a. What is the mass of a binary system with an orbital period of 4 hours and an average separation of 0.005 AU?
   b. Is this system a good candidate to contain a black hole?


**PRACTICE TEST**
**Multiple Choice Questions**

1. _____ describes pulsars as rotating neutron stars with strong magnetic fields that confine high speed charged particles in two beams emanating from the magnetic poles of the neutron star.
   a. Time dilation
   b. The lighthouse model
   c. The gravitational red shift
   d. The Chandrasekhar theory
   e. none of the above

2. The _____ of a black hole in a binary system can be very hot and emit X-rays.
   a. accretion disk
   b. event horizon
   c. Schwarzschild radius
   d. singularity
   e. magnetic field

3. The density of a _____ is less than the density of a _____.
   a. pulsar          neutron star
   b. black hole      neutron star
   c. white dwarf     red dwarf
   d. white dwarf     pulsar
   e. black hole      white dwarf

4. Current theory predicts that a neutron star cannot have a mass greater than about
   a. 0.08 solar masses.
   b. 1.4 solar masses.
   c. 3 solar masses.
   d. 10 solar masses.
   e. 100 solar masses.

5. Time dilation, as defined in general relativity, says that
   a. light inside the event horizon cannot escape the gravitational field of a black hole.
   b. neutron stars should slow down as they grow older.
   c. neutron stars should spin faster when matter is dumped on them.
   d. accretion disks around black holes should emit X-rays.
   e. time runs at a slower rate in large gravitational fields.

6. The short duration of the pulses from a pulsar tell us
   a. that the pulsar must have a strong magnetic field.
   b. that the pulsar must change its radius by only a small amount.
   c. that the pulsar must have a very small radius.
   d. that pulsars cannot have accretion disks.
   e. that pulsars must have accretion disks.

7. _____ has a radius equal to the Schwarzschild radius.
   a. An accretion disk
   b. A supernova remnant
   c. A main sequence star
   d. A pulsar
   e. The event horizon

8. Although neutron stars are very hot, they are not easy to locate because
   a. light does not escape from their event horizon.
   b. they have small surface areas.
   c. solid neutron material cannot radiate photons.
   d. they are only found in other galaxies.
   e. most lie beyond dense dust clouds.

9. A hypernove is a theory that suggests that
   a. a neutron star will collapse if it accretes more than 3 solar masses onto its surface.
   b. the most massive stars smother their supernova explosions, producing gamma ray bursters.
   c. a white dwarf will collapse to a neutron star when it accretes more than 1.4 solar masses.
   d. pulsar flashes occur as small regions of hydrogen fuse on the surface of a neutron star.
   e. time dilation causes black holes to evaporate and produce large quantities of gamma rays.

10. The Schwarzschild radius of a 7 solar mass black hole is approximately
    a. 2.3 km.
    b. 1.9 km.
    c. 21 km.
    d. 49 km.
    e. 343 km.

**Fill in the Blank Questions**

11. If the core of a collapsing star has a mass greater than _____, there is no known force that can stop its collapse.

12. A _____ _____ _____ occurs as the light traveling out from a strong gravitational field loses energy and shifts the wavelength of the light toward longer wavelengths.

13. To determine if the compact object in an X-ray binary is a neutron star or a black hole, we must determine the _____ of the compact object.

14. The _____ _____ has a radius equal to the Schwarzschild radius.

15. We expect neutron stars to spin rapidly because they conserve _____ _____.

**True-False Questions**

16. The escape velocity of an star depends on the mass of that star and the distance from the center of the star and the escaping object.

17. If the sun became a black hole, Earth would be drawn in by the black hole's huge gravitational field.

18. The density of a neutron star is about the same as that of a white dwarf.

19. It is difficult to commit fraud in science because scientific journals only allow certain highly trusted individuals to publish their work.

20. The orbit of the binary pulsar studied by Taylor and Hulse is growing smaller, presumably by emitting gravitational waves.


## ADDITIONAL READING

Bernstein, Jeremy "The Reluctant Father of Black Holes." *Scientific American 277* (June 1996), p. 80.

Comins, Neil F. "Get the Hole Story." *Astronomy 29* (April 2001), p. 48.

Lasota, Jean-Pierre "Unmasking Black Holes." *Scientific American 280* (May 1999), p. 40.

Nadis, Steve "Hunting for the Strangest Matter." *Astronomy 28* (April 2000), p. 56.

Smith, Myron "The X-Rays from Cassiopeia's Lap." *Mercury 27* (May/June 1998), p. 26.

Susskind, Leonard "Black Holes and the Information Paradox." *Scientific American 278* (April 1997), p. 52.

## ANSWERS TO MORE MATH RELATED PROBLEMS
1. 37,700 km/hr = 10.5 km/s
2. 6,100 km/hr = 1.7 km/s
3. 3 million km
4. 24 km
5. 4.3 solar masses
6. a. 0.6 solar masses
   b. no the mass is small enough that the unseen companion could be a red dwarf, white dwarf, neutron star, or black hole.


## ANSWERS TO PRACTICE TEST

| | | |
|---|---|---|
| 1. b | 8. b | 15. angular momentum |
| 2. a | 9. b | 16. T |
| 3. d | 10. c | 17. F |
| 4. c | 11. 3 solar masses | 18. F |
| 5. e | 12. gravitational red shift | 19. F |
| 6. c | 13. mass | 20. T |
| 7. e | 14. event horizon | |

# C H A P T E R   1 2

# THE MILKY WAY GALAXY

## UNDERSTANDING THE CONCEPTS

**12-1   The Discovery of the Galaxy**
**The Great Star System**
What shape did astronomers in the 1700's find our galaxy to be?
What is the central concept of this sub-section?
**Cepheid Variable Stars**
Where in the H-R diagram is the instability strip located?
What causes Cepheids to pulsate?
What is the period-luminosity relation?
What is the central concept of this sub-section?
**The Size of the Milky Way**
What did Shapley notice about the distribution of globular clusters?
How did Shapley improve upon what Leavitt had done?
What did Shapley discover about the location of the sun in our galaxy?
Why weren't astronomers before Shapley able to determine that the sun was not at the center of the galaxy?
What is the central concept of this sub-section?
**Window on Science 12-1.  The Calibration of One Parameter to Find Another**
What does it mean to calibrate a certain process?
What is the central concept of this sub-section?
**An Analysis of the Galaxy**
What are the two primary components of our galaxy?
Where does star formation take place in our galaxy?
How far is the sun from the center of the galaxy?
What are the two major parts of the spherical component of our galaxy?
What is the halo of our galaxy?
What is the nuclear bulge of our galaxy?
What is the central concept of this sub-section?
**The Mass of the Galaxy**
How can we find the mass of our galaxy?
What is differential rotation?
What is a rotation curve and what does it tell us about our galaxy?
What fraction of our galaxy seems to be composed of dark matter?
What is the central concept of this sub-section?

What are the fundamental ideas presented in this section?

**12-2   The Origin of the Milky Way**
**The Age of the Milky Way**
What is the estimated age of the disk of our galaxy?
What is the estimated age of the halo of our galaxy?
What are some of the problems encountered in determining these ages?
What is the central concept of this sub-section?

## Stellar Populations
What is a population I star?

What is a population II star?

What do astronomers mean when they talk about metals in a star?

What are two differences between the population I and population II stars?

What is the central concept of this sub-section?

## Window on Science 12-2. Understanding Nature as Processes
Why is it important to recognize processes in science?

What is the central concept of this sub-section?

## The Element-Building Cycle
How are atoms more massive than iron atoms made?

What elements should have been in our early galaxy?

Based on the process of element building, how are population I stars related to population II stars?

What is the central concept of this sub-section?

## The History of the Milky Way Galaxy
According to the traditional hypothesis for the formation of the galaxy, how did the halo and disk form?

Why does the range in age of the globular clusters create a problem for the traditional hypothesis for the formation of our galaxy?

How might have interactions between our galaxy and other galaxies affected the ages of the globular clusters?

What is the central concept of this sub-section?

What are the fundamental ideas presented in this section?

## 12-3 Spiral Arms and Star Formation
### Tracing the Spiral Arms
Why are O and B associations good spiral tracers?

Why can we conclude that O and B stars must have formed in the spiral arms where we observe them?

What is the central concept of this sub-section?

### Radio Maps of Spiral Arms
Where in the galaxy is the cool hydrogen that produces 21-cm radiation found?

Where in the galaxy are the giant molecular clouds that radiate CO emission found?

Why do we believe that the spiral arms are sites of star formation in our galaxy?

What is the central concept of this sub-section?

### The Density Wave Theory
What feature in our galaxy does the density wave theory seek to describe?

How many spiral arms do Mathematical models based on the density wave theory always produce?

What are the two problems with the density wave theory?

What is the central concept of this sub-section?

### Star Formation in Spiral Arms
What is self-sustaining star formation?

What is one limitation of self-sustaining star formation in replicating the observations of spiral arms?

What is the central concept of this sub-section?

What are the fundamental ideas presented in this section?

**12-4  The Nucleus**
    **The Center of the Galaxy and Sagittarius A* on pages 250 and 251**
        What is Sagittarius A*?
        What is believed to lie at the very center of our galaxy?
        What is the central concept of this sub-section?

    What are the fundamental ideas presented in this section?


## KEY CONCEPTS

Astronomy attempts to determine how celestial objects evolved into the forms that we see today. In the case of our galaxy, the information we currently have is not sufficient to allow us to construct a good model of how our galaxy formed. We are in the process of determining the structure and important physical phenomena that shape the structure and control the dynamics of our galaxy.

The most important sections of the chapter are those on nucleosynthesis, the spiral arms, and the nucleus. The discussion of nucleosynthesis provides a direct application of the last three chapters to develop an understanding of a much larger problem. In studying stellar evolution we learned that massive stars can generate nearly all of the elements with atomic masses less than that of iron in their cores through nuclear fusion. The only process we know of to create the elements with masses greater than iron is a supernova. As we survey the galaxy we find that the stars of the spherical component contain a much smaller relative abundance of elements more massive than helium, referred to by astronomers as metals, than do the stars of the disk component. This indicates that the material that formed the spherical component is much less processed, and hence more representative of the material from which our galaxy originally formed. The stars of the disk component have a relatively high abundance of metals indicating that they formed from material that had been processed through several supernova events. This evidence starts to give us a picture of how the galaxy may have formed.

Spiral arms have always been one of the most fascinating aspects of galaxies. The spiral arms allow us to use the information from Chapters 6, 8, and 9 to explore the structure of the spiral arms and to investigate their significance. An understanding of spiral arms is important in understanding the morphology of distant galaxies discussed in Chapter 13. The spiral arms are the regions where the HI and molecular clouds are thick and star formation is active. Realizing that star formation occurs primarily in the spiral arms of the disk of the galaxy has lead to the density wave theory to describe the initiation of star formation. This mechanism that produces the density waves is not known. We will find that many other spiral galaxies exist and that most of them appear to be related to galaxy mergers or collisions. The spiral pattern may well be the result of interactions between neighboring galaxies. As pointed out, the Milky Way has at least two satellite galaxies with which it is currently interacting.

Finally the nucleus of our galaxy is just beginning to bare itself to us. The processes we are finding will help us understand some of processes associated with active galaxies encountered in Chapter 14. Sgr A* is a phenomenal object that requires observations at as many wavelengths as possible to be able to begin to understand its behavior and importance. Sgr A* can be observed in X-ray, infrared, and radio wavelengths. It shows no movement, has a mass of about 3 million solar masses, and is about 4 AU in diameter. As we will see in Chapter 14 many other large galaxies have very compact energetic centers that appear to be black holes of a few million solar masses.


## WEBSITES OF INTEREST

    http://seds.lpl.arizona.edu/                           Messier catalog
    http://www.aao.gov.au/images.html                Anglo-Australian Observatory.
    http://www.damtp.cam.ac.uk/user/gr/public/gal_home.html    Cambridge Galaxies and Cosmology

## QUESTIONS ON CONCEPTS

1.  Why are Cepheid variable stars important in our study of the Milky Way galaxy?

    Solution:
    The Cepheid variables have a well calibrated period-luminosity relation. This means that the period of a Cepheid's intensity variations is related to its luminosity in a manner that we understand. When we measure the period of the intensity variations, we can determine the luminosity and absolute magnitude directly using the relation plotted in Figure 12-4. When we locate Cepheid variables in star clusters we can measure their periods and apparent magnitudes. We can then calculate the distance to the cluster. This gives us a way to map the distribution of the globular clusters. The Cepheids give us a reliable way to determine distance within our galaxy. Cepheids also can be found in nearby galaxy and have allowed us to determine the distances to nearby galaxies as well.

2.  Based on the process of element building, how are population I stars related to population II stars?

    Solution:
    Population I star contain much greater abundances of metals, which means that the material in them has been processed through several supernovae. The Population II stars have relatively few metals and indicated that the material in them is not as highly processed. The population II stars we see are all low to moderate mass stars, incapable of forming supernovae. However, it is their high mass counterparts that did supernova to produce the material out of which population I stars eventually formed. In short, population II stars represent the survivors of the generation of stars that produced the subsequent generations that lead to the population I stars.

3.  Why does the rotation of our galaxy suggest that it is more massive than previously thought?

    Solution:
    The rotation curve of our galaxy indicates that the mass of the galaxy is much larger than is indicated by the light it produces. In particular, the light curve tells us that a large amount of material is located in the outer reaches of our galaxy, beyond the halo. This matter does not produce light and has been dubbed dark matter. If most of the mass of the galaxy was in the form of stars, nebula and gas and dust seen in the disk, the rotation velocity of stars should decrease with increasing distance from the center of the galaxy. We would expect objects near the galactic center to orbit fairly rapidly, and objects near the outer edge of the galaxy to have very slow orbital velocities. Instead we find that the orbital velocities are nearly constant regardless of distance from the center. This indicates that the mass increases greatly as you mover further and further out into the galaxy. This is not what we see from the distribution of stars and gas in the galaxy. Therefore, the rotation curve indicates that most of the mass of the galaxy is not in the form of visible matter and much of it is in the outer reaches of the galaxy.

## WORKED EXAMPLES OF PROBLEMS REQUIRING MATHEMATICS

1.  If a globular cluster contains an RR Lyrae variable with an apparent magnitude of 16, how far away is the cluster.

    Solution:
    Figure 12-4 indicates that the absolute visual magnitudes of all RR Lyrae variables are between +1.0 and 0.0. So lets assume that the absolute visual magnitude of the star is +0.5.

    Now we know:
    the absolute visual magnitude = $M_v$ = +0.5 and
    the apparent visual magnitude = $m_v$ = 16.

We can calculate the distance to the star and cluster using the formula in *By the Numbers 8-2* on page 137.

$$d(\text{pc}) = 10^{(m_v - M_v + 5)/5} = 10^{(16 - 0.5 + 5)/5} = 10^{(20.5)/5} = 10^{4.1} = 12{,}600\,\text{pc}$$

The distance to the cluster is 12,600 pc = 12.6 kpc.

2. The apparent visual magnitude of δ Cephei is 4.0, what is its distance? (Hint use Figure 12-4 to obtain the absolute visual magnitude.)

Solution:
Figure 12-4 shows us that δ Cephei has an absolute visual magnitude = $M_v = -3.3$.
We also know the apparent visual magnitude = $m_v = 4.0$.
We can use the equation in *By the Numbers 8-2* on page 137 to find the distance to δ Cephei.

$$d(\text{pc}) = 10^{(m - M + 5)/5} = 10^{(4.0 - (-3.3) + 5)5} = 10^{(12.3)/5} = 10^{2.46} = 290\,\text{pc}$$

δ Cephei is at a distance of about 290 parsecs.

3. If the distance to the center of our galaxy is 8.5 kpc and the sun orbits the galaxy at 220 km/s, what is the minimum mass of our galaxy?

Solution:
We know
the distance to the center of our galaxy = $r$ = 8.5 kpc = 8,500 pc and
the orbital velocity = $V_c$ = 220 km/s.

We will use the equation from *By the Numbers 8-4* on page 143.

$$M(\text{solar masses}) = \frac{r_{AU}^3}{P_{\text{years}}^2}$$

$r_{AU}$ = the orbital radius in AU = 8.5 kpc = 8,500 pc = (8,500 pc)·(206,265 AU/pc) = $1.8 \times 10^9$ AU,
$P$ = the period of the orbit in years = something we need to work out, and
$M$ = the mass of the galaxy in solar masses.

We need to convert the orbital distance from parsecs to Aus. In Table A-6 on page 456 we find that 1 pc = 206,265 AU. With this conversion factor we find;
$r_{AU}$ = the orbital radius in AU = 8.5 kpc = 8,500 pc = (8,500 pc)·(206,265 AU/pc) = $1.8 \times 10^9$ AU,

New we need to find the period of the orbit. The period is equal to the circumference of the orbit divided by the orbital speed.

$$P = \frac{circumfernce}{orbital\ speed} = \frac{2\pi r}{V} = \frac{2\pi \cdot \left(1.8 \times 10^9\,\text{AU}\right) \cdot \left(1.5 \times 10^8\,\frac{\text{km}}{\text{AU}}\right)}{220\frac{\text{km}}{\text{s}}} = \frac{1.70 \times 10^{18}\,\text{km}}{220\frac{\text{km}}{\text{s}}} = 7.7 \times 10^{15}\,\text{s}$$

$$P = \left(7.7 \times 10^{15}\,\text{s}\right) \cdot \left( \frac{1\ \text{year}}{\left(60\,\frac{\text{s}}{\text{min}}\right) \cdot \left(60\,\frac{\text{min}}{\text{hour}}\right) \cdot \left(24\,\frac{\text{hours}}{\text{day}}\right) \cdot \left(365.25\,\frac{\text{days}}{\text{year}}\right)} \right)$$

$$P = \left(7.7 \times 10^{15} \text{ s}\right) \cdot \left(\frac{1 \text{ year}}{\left(3.16 \times 10^7 \frac{\text{s}}{\text{year}}\right)}\right) = 2.4 \times 10^8 \text{ years}$$

Now we are ready to solve the problem.

$$M \text{ (solar masses)} = \frac{r_{AU}^3}{P_{years}^2} = \frac{\left(1.8 \times 10^9 \text{ AU}\right)^3}{\left(2.4 \times 10^8 \text{ years}\right)^2} = 1.0 \times 10^{11} \text{ solar masses}$$

So the mass of the galaxy is $1.0 \times 10^{11}$ solar masses.

4.  If we assume that a globular cluster is 20 pc in diameter, how far away is the globular cluster if its angular diameter is 2.5 minutes of arc?

Solution:
We know
*linear diameter* = 20 pc and
*angular diameter* = 5.5 minutes of arc = 5.5×60 seconds of arc = 330 seconds of arc.

We can use the small angle formula of *By the Numbers 3-1* on page 32 to find the distance.

$$\frac{angular\ diameter}{206{,}265"} = \frac{linear\ diameter}{distance}$$

We want to solve for the distance so first we invert both sides of the equation.

$$\frac{206{,}265"}{angular\ diameter} = \frac{distance}{linear\ diameter}$$

Now we multiply both sides by the *linear diameter*.

$$linear\ diameter \cdot \frac{206{,}265"}{angular\ diameter} = linear\ diameter \cdot \frac{distance}{linear\ diameter} = distance$$

Rearranging the equation and substituting the known values we find the *distance*.

$$distance = linear\ diameter \cdot \frac{206{,}265"}{angular\ diameter} = \left(20\,\text{pc}\right) \cdot \left(\frac{206{,}265"}{330"}\right) = 12{,}500 \text{ pc} = 12.5 \text{ kpc}$$

The distance to the globular cluster would be 12.5 kpc.

## MORE MATH RELATED PROBLEMS
(Answers at the end of the chapter)

1. What is the distance to a cluster that contains an RR Lyrae variable with an apparent visual magnitude of 18.3?

2. If a Type I Cepheid with a period of 10 days and an apparent visual magnitude of 9.3 is found in a cluster, what is the distance to that cluster?

3. A cluster is found to be at a distance of 9 kpc and has an angular diameter of 3.4 minutes of arc. What is the linear diameter of this cluster?

4. A cluster contains a type II Cepheid with a period of 3 days and an apparent visual magnitude of 14. The cluster also has an angular diameter of 5.3 minutes of arc.
   a. What is the distance to this cluster?
   b. What is the linear diameter of this cluster?

5. Objects located 300 AU from the center of our galaxy are found to obit the center of the galaxy in 3.0 years. Based on this information, what is the mass of the material located within 300 AU of the center of our galaxy?

6. We measure the orbital speed of a distant star to be 220 km/s relative to the center of the galaxy. The star is located 10.6 kpc from the center of the galaxy.
   a. How long does it take this object to orbit the galaxy?
   b. What is the minimum mass of the galaxy based on this stars orbit?


## PRACTICE TEST
## Multiple Choice Questions

1. The rotation curve of the galaxy provides evidence for the existence of
   a. black holes in the cores of most globular clusters.
   b. the nuclear bulge.
   c. the galactic corona.
   d. large complexes of neutral hydrogen.
   e. the spiral arms in our galaxy.

2. Which of the following are characteristics of the stars of the disk component of our galaxy?
   I. circular orbits
   II. randomly inclined orbits
   III. higher metal abundance
   IV. ages less than 10 billion years

   a. I & II
   b. II & III
   c. III & IV
   d. I, III, & IV
   e. I, II, III, & IV

3. The traditional theory states that the galaxy formed
   a. from a large cloud of material that broke off a larger galaxy.
   b. from material that had been ejected in the violent explosion of a dying galaxy.
   c. as a result of mergers between several smaller groups of gas, dust, and stars.
   d. as two massive galaxies collided.
   e. as a large spherical cloud of gas that was rotating very slowly.

4.  The average chemical composition of stars now forming differs from that of older stars because
    a.  the material in the disk of the galaxy has fewer metals than when the galaxy formed.
    b.  supernova explosions have increased the abundance of metals in the galaxy.
    c.  the older stars used up most of the hydrogen and helium, leaving only metals for the new stars.
    d.  the old stars were not formed in our galaxy.
    e.  the new stars are made of a larger fraction of dark matter than the older stars.

5.  Infrared radiation from the galactic center
    a.  comes almost entirely from neutral hydrogen.
    b.  tells us that the stars near the center of our galaxy are extremely crowded.
    c.  shows that Sgr A$^*$ shows a small, but perceptible wobble.
    d.  shows us that most of the stars there are population I objects.
    e.  cannot penetrate the gas and dust of the galactic disk.

6.  The spherical component of our galaxy
    a.  contains predominantly population I stars.
    b.  contains large HI regions.
    c.  includes the disk and spiral arms.
    d.  includes the halo and nuclear bulge.
    e.  is on average younger than the disk component.

7.  Elements with masses less than the mass of iron and heavier than helium
    a.  were produced by nucleosynthesis during supernova explosions.
    b.  existed before the first stars formed.
    c.  were produced primarily through nucleosynthesis inside stars.
    d.  occur in large amounts in population II stars.
    e.  occur in large amounts in the stars found in the nuclear bulge.

8.  The energy source at the center of our galaxy
    I.   is not visible at optical wavelengths.
    II.  produces X-rays.
    III. is than 10 AU in diameter.
    IV.  suggests that it is powered by a 3 million solar mass black hole.

    a.  I, II, & III
    b.  I, III, & IV
    c.  I, II, & IV
    d.  II, III, & IV
    e.  I, II, III, & IV

9.  Comparing the ages of the globular cluster stars to the ages of open clusters tells us that the
    a.  masses of stars in globular clusters is much larger than those in open clusters.
    b.  disk and spherical components of the galaxy are roughly the same age.
    c.  spherical component is older than the disk component of the galaxy.
    d.  spherical component is younger than the disk component of the galaxy.
    e.  spherical component is composed of a later generation of stars than the disk component.

10. A Type II Cepheid with an absolute visual magnitude of –2 and an apparent visual magnitude of 13 is located in a globular cluster.  What is the distance to this cluster?
    a.  10 kpc
    b.  10 pc
    c.  100 kpc
    d.  100 pc
    e.  1 pc

**Fill in the Blank Questions**

11. _____ stars have relatively large amounts of metals.

12. The _____ component of the galaxy contains population II stars and globular clusters, and very little gas and dust.

13. The disk of the Milky Way is approximately _____ ly in diameter.

14. Spectral type _____ and _____ stars are good tracers of spiral arms because they are very luminous.

15. 21 cm radiation shows us the locations of _____ clouds.

**True-False Questions**

16. The center of our galaxy is located in the direction of the constellation of Sagittarius.

17. Our galaxy is suspected to be surrounded by a galactic corona because the disk of the galaxy rotates faster than expected in its outer region.

18. Younger stars have more heavy elements because old stars destroy heavy elements as they age.

19. The traditional theory of the formation of our galaxy cannot explain the difference in metal abundance of the population I and II stars.

20. The density wave theory accurately predicts the spurs and branches of the spiral arms in our galaxy?

**ADDITIONAL READING**

Bartusiak, Marcia "A Beast in the Core." *Astronomy 26* (July 1998), p. 48.

Eicher, David "Galactic Genesis." *Astronomy 27* (May 1999), p. 38.

Kaiser, Denise "Cosmic Intrigue." *Astronomy 28* (Oct. 2000), p. 42.

Szpir, Michael "Passing the Bar Exam." *Astronomy 27* (March 1999), p. 46.

**ANSWERS TO MORE MATH RELATED PROBLEMS**
1. 40,000 pc = 40 kpc
2. 7.2 kpc, taking the absolute visual magnitude to be –5.0
3. 8.9 pc
4. a. 10 kpc, taking the absolute visual magnitude to be –1.0
   b. 15.4 pc
5. 3,000,000 solar masses = $3.0 \times 10^6$ solar masses
6. a. $3.0 \times 10^8$ years
   b. $1.2 \times 10^{11}$ solar masses

**ANSWERS TO PRACTICE TEST**

1. c
2. d
3. e
4. b
5. b
6. d
7. c

8. e
9. c
10. a
11. Population I
12. spherical
13. 75,000
14. O and B

15. neutral hydrogen or HI
16. T
17. T
18. F
19. F
20. F

# CHAPTER 13

# GALAXIES

## UNDERSTANDING THE CONCEPTS

### 13-1 The Family of Galaxies
**The Shapes of Galaxies and Galaxy Classification on pages 258 and 259**
How does the presence of gas and dust influence the appearance of galaxies?
Upon what property is the Hubble classification of galaxies based?
What are the three basic types of galaxies?
What are the basic properties of an elliptical galaxy?
What is the central concept of this sub-section?
**Window on Science 13-1. Searching for Clues Through Classification in Science**
What benefit is classification in science?
What is the central concept of this sub-section?
**How Many Galaxies**
What is the most common type of galaxy?
Why do we see more spiral galaxies than elliptical, yet believe that there are actually more elliptical galaxies than spiral galaxies?
What is the central concept of this sub-section?
**Window on Science 13-2. Selection Effects in Science**
What is a selection effect?
Why are selection effects especially difficult for astronomers to avoid?
What is the central concept of this sub-section?

What are the fundamental ideas presented in this section?

### 13-2 Measuring the Properties of Galaxies
**Distance**
What are distance indicators?
What are some distance indicators astronomers use?
How are distance indicators calibrated?
What is meant by the "look-back time"?
What is the central concept of this sub-section?
**The Hubble Law**
The Hubble law is a relationship between which two observable properties of a galaxy?
What is the current best value for the Hubble constant, including the units?
What is the central concept of this sub-section?
**Diameter and Luminosity**
What is the key to determining a galaxies diameter and luminosity?
Are the diameter and luminosity of a galaxy related to its Hubble classification?
What is the central concept of this sub-section?
**Mass**
What are two methods astronomers can use to determine the mass of a galaxy?
What is the range in mass for galaxies?
What is the central concept of this sub-section?

### Supermassive Black Holes in Galaxies

Why do astronomers conclude that the centers of many galaxies contain supermassive black holes?

The mass of the black hole at the center of a galaxy appears to be related to the mass of which part of the galaxy?

What is the central concept of this sub-section?

### Dark Matter in Galaxies

Where is most of the dark matter in a galaxy located?

What fraction of the mass of the universe is dark matter?

What are some suggestions for what the dark matter might be?

What is the central concept of this sub-section?

What are the fundamental ideas presented in this section?

## 13-3   The Evolution of Galaxies

### Clusters of Galaxies

What are the characteristics of a rich galaxy cluster?

What are the characteristics of a poor galaxy cluster?

What is the Local Group?

What is the central concept of this sub-section?

### Colliding Galaxies

Why do we expect galaxies to collide relatively frequently?

When galaxies collide or interact, what often happens to the rate of star formation in the galaxies?

What is galactic cannibalism?

What three galaxies is the Milky Way cannibalizing?

What is the central concept of this sub-section?

### The Origin and Evolution of Galaxies

What phenomena play an important role in the evolution of most galaxies?

What is a starburst galaxy?

In what ways are the galaxies at great look-back times different than galaxies are today?

What is the central concept of this sub-section?

What are the fundamental ideas presented in this section?

## KEY CONCEPTS

The real point of this chapter is that the study of galaxies is in its infancy and a lot of progress is still to be made.  Yet, the observations we have allow us to construct and test theories that begin to describe galaxy evolution.

The first section emphasizes the structure of various types of galaxies.  This goes beyond simply describing their shape.  We find that the types of stars and relative abundance of gas and dust differs with different types of galaxies.  This information is the first hint at understanding the evolution of various galaxies.

The second section focuses on the measurable properties of a galaxy, which are: mass, luminosity, and diameter.  This study leads to the need for an accurately calibrated distance scale and the realization that most of the mass in the universe is in the form of dark matter.  The accurate calibration of the distance scale and an understanding of the dark matter problem are vital to understanding the ideas of cosmology presented in Chapter 15.  This section also discusses supermassive black holes located at the centers of galaxies.  These supermassive black holes will be discussed in Chapter 14 as the engine that produces the tremendous energy emitted by active galaxies.

The final section discusses the origin and evolution of galaxies and the importance of interactions between galaxies.  The study of galactic interactions in the last several years has greatly revised our views on the formation and evolution of galaxies and the universe. The size of galaxies and their separation distances are nearly the same; consequently, they should bump into each other regularly. These interactions lead to mergers and galactic cannibalism that can trigger star formation and can drastically alter the appearance of the galaxy.

Clusters of galaxies allow us to study how galaxy interactions affect galaxy evolution. Realizing that galaxies form large clusters and superclusters is important to understanding the evolution of the universe discussed in Chapter 15.

## WEBSITES OF INTEREST

| | |
|---|---|
| http://zebu.uoregon.edu/galaxy.html | Galaxies and stars |
| http://hubblesite.org/newscenter/ | Hubble Space Telescope |
| http:///www.aao.gov.au/images.html | Anglo-Australian Observatory |
| http://www.damtp.cam.ac.uk/user/gr/public/gal_home.html | Cambridge Galaxies and Cosmology |
| http://library.thinkquest.org/3461/dark.htm | Dark Matter description |

## QUESTIONS ON CONCEPTS

1.  What are the basic properties of a spiral galaxy?

    Solution:
    Spiral galaxies contain a well defined disk with spiral arms. These galaxies possess relatively large amounts of gas and dust. The spiral arms appear blue and show bright O and B stars as well as emission nebula indicating that star formation is occurring in the disks of these galaxies. The nuclear bulges appear yellow with little gas or dust, indicating that star formation is not occurring in the nuclear bulge. The spirals also tend to be larger in diameter and more luminous than the average for all galaxy types.

2.  How are distance indicators calibrated?

    Solution:
    Distance indicators are calibrated in a stepwise fashion working out from well-established distance measurements. The foundation of all galactic distance indicators is the period-luminosity relation for the Cepheids. If we accept its calibration accuracy as discussed in Chapter 12, we can use it to measure the distances to several galaxies within the Local Group. Once the distances to several galaxies are known we can look for other objects whose luminosities or sizes can be calibrated. Consider the size of a globular cluster as an example. We could use Cepheid variables to determine the distance to several galaxies. If we then find that the average diameter of a globular cluster is the same regardless of the galaxy that it is in, we can use the average diameter of a globular cluster as a distance indicator. Then we can measure the angular diameters of globular clusters in a galaxy whose distance is not known and calculate the distance to the galaxy (see Worked Examples of Problems Requiring Mathematics problem number 2 below).

    Once a new distance indicator is calibrated, like the size of globular clusters discussed above, we can use them to calibrate other distance indicators like supernovae or the luminosities of spiral galaxies, etc. In this way the distance indicators are calibrated against each other to build a set of stepping stones that allow us to measure greater and greater distances.

3. Why are galaxy interactions believed to be a major process in the evolution of galaxies?

Solution:
There are several things about galaxies and the interactions of galaxies that we know. Lets write them down and then try to construct an argument to support the role interactions might play in galactic evolution.

1. Galaxies are large and fairly close together so collisions should be common.
2. Many galaxies are observed colliding and interacting with other galaxies.
3. Galaxies in collisions show an increased level of star formation.
4. The Milky Way is currently cannibalizing three galaxies, which has altered their shapes and added material to the Milky Way.

The four observations listed above indicate that galaxy interactions are fairly common. These interactions mix the material of the galaxies to alter their composition. Interactions between gas clouds in the galaxies can increase star formation. Finally, our own galaxy shows the effects of galactic interactions altering at least the shapes, structures, and dynamics of three galaxies. The effects of these galaxies on the Milky Way may be slight, but clearly the interaction is affecting the three smaller galaxies. The interaction is changing the composition of the Milky Way, even if only slightly.


## WORKED EXAMPLES OF PROBLEMS REQUIRING MATHEMATICS

1. If a galaxy contains a Type I Cepheid with a period of 25 days and an apparent magnitude of 18, what is the distance to the galaxy?

Solution:
We know
the apparent visual magnitude = $m_v$ = 18 and
the period of the Cepheid = 25 days.

We can use Figure 12-4 on page 232 to find that the absolute visual magnitude, $M_v$, of a Type I Cepheid with a 25-day period is approximately –5.5.

We can solve for the distance by using the absolute magnitude formula in *By the Numbers 8-2* on page 137.

$$d(\text{pc}) = 10^{(m_v - M_v + 5)/5} = 10^{(18 - (-5.5) + 5)/5} = 10^{(28.5)/5} = 10^{5.7} = 500,000 \,\text{pc} = 500 \,\text{kpc} = 0.5 \,\text{Mpc}$$

The galaxy is at a distance of 500 kpc = 0.5 Mpc.

2. If the average globular cluster is calibrated to be 25 pc in diameter, and a globular cluster is located in a distant galaxy with an angular diameter of 0.8 seconds of arc, how far away is the galaxy?

Solution:
We know
the linear diameter of the globular cluster = 25 pc and
the angular diameter of the cluster = 0.8 seconds of arc.

We use the small angle formula from *By the Numbers 3-1* on page 32 to solve for the distance.

$$\frac{angular\ diameter}{206,265"} = \frac{linear\ diameter}{distance}$$

We want to solve for the *distance*, so first we need to invert both sides of the equation and then multiply both sides by the *linear diameter*.

$$\frac{206,265"}{angular\ diameter} = \frac{distance}{linear\ diameter}$$

$$linear\ diameter \cdot \frac{206,265"}{angular\ diameter} = linear\ diameter \cdot \frac{distance}{linear\ diameter} = distance$$

We can rearrange this and substitute our values to find the distance.

$$distance = 206,265" \cdot \left(\frac{linear\ diameter}{angular\ diameter}\right) = 206,265" \cdot \left(\frac{25\,pc}{0.8"}\right) = 6,400,000\ pc = 6.4\ Mpc$$

The distance to the galaxy is 6,400,000 pc = 6.4 Mpc.

3.  The rotation curve of a galaxy shows us that stars at a distance of 11 kpc from the center of the galaxy have an orbital velocity of 185 km/s.
    a.  What is the orbital period of the stars at 11 kpc from the center?
    b.  What is the minimum mass of the galaxy in solar mass units?

Solution to part a:
We know
radius of the orbit = $r$ = 11.0 kpc = 11,000 pc = (11,000 pc) · (3.1×10$^{13}$ km/pc) = 3.4×10$^{17}$ km and the orbital speed = $v$ = 185 km/s.

The period of the orbit is equal to the distance traveled divided by the speed. In this case the distance traveled is the circumference of the orbit = $2\pi r$ = $2\pi$ (3.4×10$^{17}$ km) = 2.1×10$^{18}$ km.

Then

$$P = \frac{circumference}{orbital\ speed} = \frac{2.1\times10^{18}\ km}{185\ km/s} = 1.2\times10^{16}\ s = \frac{1.2\times10^{16}\ s}{3.2\times10^{7}\ \frac{s}{year}} = 3.7\times10^{8}\ years$$

The orbital period is 3.7×10$^{8}$ years (370 million years).

Solution to part b:
We know
the orbital period = P = 3.7×10$^{8}$ years, and
the orbital distance = $r$ = 11.0 kpc = 11,000 pc×206,265 AU/pc = 2.3×10$^{9}$ AU.

We can find the mass of the galaxy using the orbital formula from *By the Numbers 8-4* on page 143. This formula requires that the period be given in years and that the distance be given in AU. The mass will be in solar mass units.

$$M = \frac{r_{AU}^{3}}{P_{years}^{2}} = \frac{\left(2.3\times10^{9}\ AU\right)^{3}}{\left(3.7\times10^{8}\ years\right)^{2}} = \frac{1.2\times10^{28}}{1.4\times10^{17}} = 8.6\times10^{10}\ solar\ masses$$

The mass of the galaxy is at least 8.6×10$^{10}$ solar masses (86 billion solar masses).

4. If a galaxy has a radial velocity of 4,300 km/s and the Hubble constant is 70 km/s/Mpc, how far away is the galaxy?

Solution:
We know
the recessional velocity = $V_r$ = 4,300 km/s and
the Hubble constant = $H$ = 70 km/s/Mpc.

We use the Hubble law in *By the Numbers 13-1* on page 263.

$$V_r = H \cdot d$$

If we divide both sides by $H$ we can solve for the distance, $d$.

$$\frac{V_r}{H} = \frac{H \cdot d}{H} \quad \text{so} \quad d = \frac{V_r}{H} = \frac{4,300 \text{ km/s}}{70 \frac{\text{km/s}}{\text{Mpc}}} = 61 \text{ Mpc}.$$

The distance to the galaxy is 61 Mpc.

## MORE MATH RELATED PROBLEMS
(Answers at the end of the chapter)

1. What is the distance to a galaxy with a recessional velocity of 15,000 km/s if the Hubble constant is 70 km/s/Mpc?

2. Assume the Hubble constant is 70 km/s/Mpc and complete the following table.

| Galaxy | Recessional velocity (km/s) | Distance (Mpc) |
|---|---|---|
| A | 1,340 | |
| B | 3,100 | |
| C | | 15.0 |
| D | | 3.7 |
| E | 400 | |

3. A globular cluster with an angular diameter of 4.5 seconds of arc is located in a galaxy. If the average linear diameter of a globular cluster is 25 pc, what is the distance to this galaxy?

4. If the smallest angular diameter we can measure is 0.1 seconds of arc, and the average globular cluster is 25 pc in diameter, what is the maximum distance we can measure using the diameters of globular clusters?

5. A galaxy is found with stars orbiting with a period of 25 million years. It is determined that these stars are 4.3 kpc from the center of the galaxy. What is the mass of the galaxy?

6. Stars in a galaxy have an orbital velocity of 325 km/s and are located at a distance from the center of the galaxy of 6.8 kpc.
   a. What is the orbital period of these stars?
   b. What is the mass of the galaxy?

**PRACTICE TEST**
**Multiple Choice Questions**

1. The Hubble law is a relationship between which two observable properties of a galaxy?
   a. distance and velocity of recession
   b. distance and mass
   c. mass and luminosity
   d. luminosity and distance
   e. mass and diameter

2. The orbital motion of material near the center of a galaxy can be used to determine
   a. the luminosity of the galactic center.
   b. the diameter of the nuclear bulge.
   c. the age of the galaxy.
   d. the distance to the galaxy.
   e. the mass of the galactic center.

3. Why is the number of galaxies in the Local Group uncertain?
   a. The Local Group is so far away that we cannot be certain that we see the faintest galaxies.
   b. The Local Group is a rich cluster with a very dense concentration of galaxies at its center.
   c. The Local Group contains a large number of spirals which are difficult to identify
   d. Gas and dust within the Milky Way obscure our view of several regions of the Local Group.
   e. The Local Group contains a large number of colliding galaxies.

4. Spiral galaxies
   a. have blue tinted disks and yellow nuclear bulges.
   b. have yellow disks and blue tinted nuclear bulges.
   c. appear to form in isolation and are seldom found in poor clusters.
   d. appear to form primarily at the centers of rich clusters.
   e. seldom contain a supermassive black hole at their center.

5. An elliptical galaxy might
   a. evolve into an irregular galaxy when it has used up all of its gas and dust.
   b. evolve from a single spiral galaxy when the spiral has used up all of its gas and dust.
   c. become a starburst galaxy if it were to move through the hot intergalactic medium of a cluster.
   d. be formed from the collision and merger of spiral galaxies.
   e. evolve from an S0 galaxy if the S0 galaxy were to increase its rotation rate.

6. Starburst galaxies
   a. are generally dwarf elliptical galaxies.
   b. contain a large number of very old stars and almost no gas or dust.
   c. are often associated with a galaxy that is colliding with another galaxy.
   d. are only found at very large look-back times.
   e. are composed of filaments and voids.

7. The _____ method of finding the mass of a galaxy depends on the motions of galaxies within a cluster of galaxies.
   a. cluster
   b. rotation
   c. binary
   d. parallax
   e. elliptical

8. The Milky Way galaxy is
   a. part of a rich cluster known as the Virgo cluster.
   b. part of a rich cluster known as the Local Group.
   c. part of a poor cluster known as the Virgo Cluster.
   d. part of a poor cluster known as the Local Group.
   e. an isolated galaxy that isn't part of a galaxy cluster.

9. If the absolute magnitude of a supernova is −19 and a galaxy is found that contains a supernova with an apparent magnitude of 21, what is the distance to the galaxy?
   a. 1000 AU.
   b. 1000 pc.
   c. 1000 ly.
   d. 1000 kpc.
   e. 1000 Mpc.

10. If $H$ equals 70 km/sec/Mpc, then a galaxy with a radial velocity of 2800 km/sec has a distance of approximately
   a. 2870 Mpc
   b. 2730 Mpc
   c. 40 Mpc
   d. 0.025 Mpc
   e. 196,000 Mpc

**Fill in the Blank Questions**

11. The _____ _____ is a relationship between the distance to a galaxy and the galaxy's recessional velocity.

12. _____ clusters of galaxies contain less than 1000 galaxies and seldom contain giant elliptical galaxies.

13. Based on the galaxies found in the Local Group of galaxies, the most common type of galaxy in the universe is expected to be the _____ galaxies.

14. More than 90% of the mass of a galaxy is composed of _____.

15. _____ galaxies contain a well defined disk, upper and lower main sequence stars and gas and dust.

**True-False Questions**

16. We should expect galaxies to collide fairly often because they are relatively large with respect to their separation distances.

17. The rotation curve of a galaxy can be used to determine the luminosity of the galaxy.

18. It is believed that ring galaxies form when two galaxies collide nearly head-on at high speed.

19. Observations of galaxies and clusters of galaxies indicate that about 95 per cent of the universe is dark matter.

20. Galactic cannibalism refers to the destruction of a galaxy's globular clusters by the galaxy's nucleus.

## ADDITIONAL READING

Conti, Peter S. "Bursting Onto the Scene." *Mercury 26* (May/June 1997), p.

Graham, David "Clusters in Collisions." *Astronomy 27* (May 1999), p. 58.

Martin, Pierre, and Daniel Friedli "At the Hearts of Barred Galaxies." *Sky and Telescope 97* (March 1999), p. 32.

Roth, Joshua "When Galaxies Collide." *Sky and Telescope 95* (March. 1998), p. 48.

## ANSWERS TO MORE MATH RELATED PROBLEMS

1.  210 Mpc
2.  Assume the Hubble constant is 70 km/s/Mpc and complete the following table

| Galaxy | Recessional velocity (km/s) | Distance (Mpc) |
|--------|------------------------------|----------------|
| A | 1,340 | **19.1** |
| B | 3,100 | **44.3** |
| C | **1,050** | 15.0 |
| D | **259** | 3.7 |
| E | 400 | **5.7** |

3.  1,100,000 pc = 1.1 Mpc
4.  52,000,000 pc = 52 Mpc
5.  $1.1 \times 10^{12}$ solar masses
6.  a.   $1.3 \times 10^{8}$ years
    b.   $1.6 \times 10^{11}$ solar masses

## ANSWERS TO PRACTICE TEST

1.  a
2.  e
3.  d
4.  a
5.  d
6.  c
7.  a
8.  d
9.  e
10. c
11. Hubble law
12. Poor
13. dwarf elliptical
14. dark matter
15. Spiral
16. T
17. F
18. T
19. T
20. F

# C H A P T E R   1 4

# GALAXIES WITH ACTIVE NUCLEI

## UNDERSTANDING THE CONCEPTS

### 14-1  Active Galactic Nuclei
#### Seyfert Galaxies
What does AGN stand for and what are AGNs?

What is a Seyfert galaxy?

What are the differences between Type 1 and Type 2 Seyfert galaxies?

Are Seyferts more common in isolated galaxies or in interacting pairs?

What is the central concept of this sub-section?

#### Double-Lobed Radio Sources and Cosmic Jets and Radio Lobes on pages 280 and 281
What are double-lobed radio sources?

How are the hot spots associated with double-lobed radio sources formed?

What are the similarities and differences between the jets from bipolar flows and active galaxies?

What is the central concept of this sub-section?

#### Window on Science 14-1.  It Wouldn't Stand Up in Court: Statistical Evidence
What is statistical evidence?

What does statistical evidence allows us to do?

What is the central concept of this sub-section?

#### Window on Science 14-2.  Scientific Arguments
Why do scientists construct scientific arguments?

What is a scientific argument?

What is the central concept of this sub-section?

#### Testing the Black Hole Hypothesis
What two properties must be observed to determine if supermassive black holes are the central energy sources of AGNs?

What is the central concept of this sub-section?

#### The Search for a Unified Model
What are the five basic structures in the unified model for active galaxies?

How does the unified model explain the observations of Type 2 Seyferts?

What is the central concept of this sub-section?

#### Black Holes and Galaxy Formation
The mass of the black hole in a galaxies center is correlated with what other feature of the galaxy?

Where have midsize black holes recently been found?

What is the central concept of this sub-section?

What are the fundamental ideas presented in this section?

### 14-2  Quasars
#### The Discovery of Quasars
The radio emission from quasars resembles the radio emission from what other objects?

What objects do quasars resemble in visible light photographs?

What sort of spectra do we obtain from quasars?

How is the Hubble law used to understand quasars?

What is the central concept of this sub-section?

### Quasar Distances
What is meant by a relativistic red shift?
Why can't we convert quasar red shifts into precise distances?
What is a gravitational lens?
How does the gravitational lensing of some quasars provide evidence that quasars are at great distances?
What is the central concept of this sub-section?

### Evidence of Quasars in Distant Galaxies
What is quasar fuzz and what sort of spectrum does it produce?
What is the central concept of this sub-section?

### A Model Quasar
What is the basic geometry that our quasar model is based?
Why is black hole necessary in the quasar model?
Where might synchrotron radiation be produced in our quasar model?
What is the central concept of this sub-section?

### Quasars Through Time
At what red shift are quasars most common?
What is our interpretation as to why quasars are most common between red shifts of 2 and 2.7?
What is the central concept of this sub-section?

What are the fundamental ideas presented in this section?


## KEY CONCEPTS

The key concept of this chapter is that all of the active galaxy classes, including quasars, fit within a unified model. The chapter stresses the evidence that was obtained and how a single model has been developed that explains each of the different classes. This chapter builds on the discussion of the previous chapter concerning the importance of collisions and mergers in the evolution of galaxies. AGNs appear much more often in interacting or distorted systems than in isolated systems, further suggesting the importance of mergers in galactic evolution.

Many astronomers accept quasars to be the most distant objects in the universe. The current evidence does support the interpretation that they are much further away and much more luminous than most normal galaxies. It is important to understand that when we observe quasars we are looking at the conditions of a galaxy at a much earlier time. Quasars tell us about the early universe and the conditions under which the first galaxies formed. Quasars stretch our understanding of the natural world to new limits and provide some interesting tests of predictions made by the theory of general relativity.


## WEBSITES OF INTEREST

http://seds.lpl.arizona.edu/messier/galaxy.html      Galaxies from the Messier catalog
http://www.damtp.cam.ac.uk/user/gr/public/gal_home.html      Cambridge Galaxies and Cosmology

## QUESTIONS ON CONCEPTS

1.  Why do we assert that galaxy interactions are important in triggering the formation of active galactic nuclei?

    Solution:
    Triggering active galactic nuclei requires two things, a very massive, ultra-compact object in the core of the galaxy, and material falling into the core that forms an accretion disk and dense torus. The orbits of stars and gas clouds in most normal galaxies are stable and circular, but the material in AGNs is flowing into the core to form the dense torus and accretion disk. Many AGNs are associated with distorted galaxies, suggesting a collision or merger, and some are observed to be involved in a collision or merger. Tidal interactions between galaxies can cause material to be both ejected from the galaxy and thrown into the center of the galaxy. Such interactions cause some of the stars and gas clouds to follow highly elliptical orbits that send the material very near the nucleus. The galactic interactions are necessary to disturb the normally stable orbits of the material and push it to move toward the core of the galaxy. Near the nucleus the density of material is very high and the stars and gas clouds interact more frequently and trap much of the material in the inner portion of the galaxy forming the dense torus. The material will be in degrading orbits and will continuously move toward the center of the galaxy, passing through the dense torus and into the accretion disk.

2.  How can quasars tell us about the environment in which early galaxies formed?

    Solution:
    Quasars are visible over an extremely large range in distances. When we observe a quasar we are looking at information from a different time period in our universe. Distance quasars, with red shifts around 6, show us something of what the universe was like 13 billion years ago, when the universe was only a billion years old. We find few quasars from this time period, and few from the current time. We find the largest number of quasars at a red shift of around 2 to 2.7. These quasars have look-back times of 11 to 12 billion years, so these quasars contain information from when the universe was 2 or 3 billion years old. Active galaxies tell us that mergers play an important role in producing the extreme energy output of active galactic nuclei and quasars. The large number of quasars at red shifts of 2 to 2.7 suggests that 11 billion to 12 billion years ago galaxy collisions were much more common. This implies that the galaxies were on average much closer together around this time.

## WORKED EXAMPLES OF PROBLEMS REQUIRING MATHEMATICS

1.  A quasar is found with a red shift of 3.5.
    a.  What is the recessional velocity of this quasar?
    b.  What is the apparent distance to this quasar if the Hubble constant is 70 km/s/Mpc?

    Solution to part a:
    We know
    the red shift of the quasar = $z = 3.5$ and
    the speed of light = $c = 3.0 \times 10^5$ km/s.

    We can use the relativistic red shift formula from *By the Numbers 14-1* on page 288 to calculate the recessional velocity.

    $$\frac{V_r}{c} = \frac{(z+1)^2 - 1}{(z+1)^2 - 1} = \frac{(3.5+1)^2 - 1}{(3.5+1)^2 + 1} = \frac{20.25 - 1}{20.25 + 1} = \frac{19.25}{21.25} = 0.91$$

    The recessional velocity of the quasar is $0.91c = 0.91 \cdot (3.0 \times 10^5 \text{ km/s}) = 2.7 \times 10^5$ km/s.

Solution to part b:
We know
the recessional velocity = $V_r = 2.7 \times 10^5$ km/s and
the Hubble constant = $H = 70$ km/s/Mpc
We can use the Hubble law from *By the Numbers 13-1* on page 263 to solve for the apparent distance.

$$V_r = H \cdot d \quad \text{If we multiply both sides by } H, \text{ we can solve for the distance, } d. \quad \frac{V_r}{H} = \frac{Hd}{H} = d$$

$$d = \frac{V_r}{H} = \frac{2.7 \times 10^5 \text{ km/s}}{70 \frac{\text{km/s}}{\text{Mpc}}} = 3{,}900 \text{ Mpc}$$

The apparent distance to the quasar is 3,900 Mpc.

2.  A radio jet is 30 seconds of arc long and is approximately 610 Mpc away.
    a.  How long is the jet?
    b.  If the gas in the jet is traveling at 5,300 km/s, how long will it take material to travel from the center of the galaxy to the end of the jet?

Solution to part a:
We know
the distance to the jet is 610 Mpc and
the angular diameter of the jet is 30 seconds of arc.

We can use the small angle formula in *By the Numbers 3-1* on page 32 to solve for the linear diameter.

$$\frac{angular\ diameter}{206{,}265"} = \frac{linear\ diameter}{distance}$$

If we multiply both sides by the distance, we can solve for the linear diameter.

$$distance \cdot \frac{angular\ diameter}{206{,}265"} = distance \cdot \frac{linear\ diameter}{distance} = linear\ diameter$$

So we can rearrange this and plug in our values.

$$linear\ diameter = distance \cdot \frac{angular\ diameter}{206{,}265"} = 610 \text{ Mpc} \cdot \frac{30"}{206{,}265"} = 0.089 \text{ Mpc} = 89 \text{ kpc}$$

The length of the jet is 0.089 Mpc or 89 kpc.

Solution to part b:
We know
the speed of the material is 5,300 km/s and
the distance to be traveled = 89 kpc = 89,000 pc = (89,000 pc) $\cdot$ (3.1$\times 10^{13}$ km/pc) = 2.8$\times 10^{18}$ km.

In this problem we want to find the time it takes to move a given distance when we know the distance and the speed. Since speed is equal to distance divided by time, the time to travel a given distance is equal to the distance divided by the speed, $t = d/v$.

$$t = \frac{distance}{speed} = \frac{2.8 \times 10^{18} \text{ km}}{5,300 \text{ km/s}} = 5.3 \times 10^{14} \text{ s} = \frac{\left(5.3 \times 10^{14} \text{ s}\right)}{\left(60 \frac{s}{min} \cdot 60 \frac{min}{hr} \cdot 24 \frac{hr}{day} \cdot 365.25 \frac{days}{year}\right)} = 1.7 \times 10^{7} \text{ years}$$

The material will take $1.7 \times 10^{7}$ years = 17 million years.

3.  A quasar has a luminosity of $3 \times 10^{51}$ J/s. How many solar masses of material would have to be converted to energy each second to supply this energy output? (Hint: 1 solar mass = $2.0 \times 10^{30}$ kg)

Solution:
We know the energy generated each second = $E = 3 \times 10^{51}$ J/s. We can use Einstein's energy formula to calculate the amount of mass that needs to be converted. Remember that this mass will be in kilograms (kg) so we will have to convert it to solar masses.

$E = mc^2$    If we divide both sides by $c^2$ we can solve for the mass.    $\frac{E}{c^2} = \frac{mc^2}{c^2} = m$

$$m = \frac{E}{c^2} = \frac{3 \times 10^{51} \text{ J/s}}{\left(3.0 \times 10^{8} \frac{m}{s}\right)^2} = 3.3 \times 10^{34} \text{ kg/s} = \frac{\left(3.3 \times 10^{34} \frac{kg}{s}\right)}{\left(2.0 \times 10^{30} \frac{kg}{\text{solar mass}}\right)} = 17,000 \frac{\text{solar masses}}{\text{second}}$$

The quasar puts out the same amount of energy as released in converting 17,000 solar masses of material to energy every second.


## MORE MATH RELATED PROBLEMS
(Answers at the end of the chapter)

1.  The red shift of B2 1208+32A is 0.389.
    a.  What is the recessional velocity of B2 1208+32A?
    b.  What is the apparent distance to B2 1208+32A if the Hubble constant is 70 km/s/Mpc?

2.  The Hα line of hydrogen is produced at a wavelength of 656.3 nm. In the quasar PKS 1217+02 the Hα line appears at a wavelength of 813.8 nm.
    a.  What is Δλ for the Hα line from PKS 1217 +02?
    b.  What is the red shift of PKS 1217 +02?
    c.  What is the recessional velocity of PKS 1217 +02?
    d.  What is the apparent distance to PKS 1217 +02, if the Hubble constant is 70 km/s/Mpc?

3.  Fill in the table below, assuming that the Hubble constant is 70 km/s/Mpc.

| Object | Red shift | $V_r/c$ | $V_r$ (km/s) | Distance (Mpc) |
|--------|-----------|---------|--------------|----------------|
| 1      | 3.2       |         |              |                |
| 2      | 0.15      |         |              |                |
| 3      | 6.5       |         |              |                |
| 4      | 2.9       |         |              |                |
| 5      | 1.8       |         |              |                |
| 6      | 0.47      |         |              |                |

4.  A quasar at 1400 Mpc shows fuzz that extends with a diameter of 2.3 seconds of arc. What is the linear diameter of the quasar fuzz?

5.  The jet of a radio galaxy has an angular length of 300 seconds of arc and the material is found to be moving at 4,300 km/s. The distance to the radio galaxy is 820 Mpc.
    a.  What is the length of the jet in km?
    b.  How long does it take material to flow from the center of the galaxy to the end of the jet?


**PRACTICE TEST**
**Multiple Choice Questions**

1.  The hot spots in radio lobes
    a.  are caused by the high temperature of the accretion disk.
    b.  are located on the inside of the lobe closest to the central galaxy.
    c.  are created as the material from the jet impacts the intergalactic medium.
    d.  rotate in the opposite direction of the galaxy.
    e.  are formed by a gravitational lens.

2.  The cores of most galaxies contain black holes, but do not produce huge amounts of radios emission, powerful jets, or excessive X-ray radiation. Why might these black holes lie dormant, while the black holes in active galaxies produce so much energy?
    a.  The black holes in active galaxies are believed to rotate faster.
    b.  The accretion disks around black holes in active galaxies are believed to contain less dust.
    c.  The accretion disks around black holes in active galaxies are believed to be dissipating.
    d.  It is believed that collisions could cause the normal galaxies to become active galaxies.
    e.  It is believed that active galaxies are really nearby quasars and unrelated to normal galaxies.

3.  What observational evidence supports the hypothesis that quasars are the centers of very distant galaxies?
    I.    Quasar fuzz has the spectra of a galaxy.
    II.   Supernovae have been found near quasars and at the same approximate distance.
    III.  Gravitational lenses reveal that quasars are very far away.
    IV.   A quasar is located at the center of the Milky Way.

    a.  I & II
    b.  III & IV
    c.  I & IV
    d.  II & III
    e.  I, II, & III

4.  The red shift of a galaxy is determined by
    a.  measuring the galaxy's apparent visual magnitude and absolute visual magnitude.
    b.  measuring the galaxy's mass and luminosity.
    c.  measuring the galaxy's rotation curve.
    d.  comparing the position of spectral lines in the galaxy's spectrum to a reference spectrum.
    e.  comparing the distance of the galaxy to the galaxy's recessional velocity.

5.  Most quasars have red shifts
    a.  between 0 and 1.2.
    b.  between 2.0 and 2.7.
    c.  between 3.2 and 3.8.
    d.  between 4.0 and 5.0.
    e.  between 6.0 and 6.5.

6. The gravitational lens effect
   I. shows that space-time is curved.
   II. produces the second lobe in double-lobe radio galaxies.
   III. provides evidence that some quasars are more distant than galaxies.
   IV. requires that a black hole be present in the galaxy that bends the light.

   a. I & II
   b. I & III
   c. II & III
   d. II & IV
   e. III & IV

7. The radio radiation emitted by a radio lobe is primarily
   a. produced by the disk of the galaxy.
   b. synchrotron radiation.
   c. the result of star formation in the thick disk.
   d. black body radiation from the cool dusty torus.
   e. observable in the ultraviolet part of the spectrum.

8. All active galaxies
   a. are at red shifts greater than 2.0.
   b. are at red shifts less than 2.0.
   c. are found near a quasar.
   d. form gravitational lenses.
   e. show a compact energetic core.

9. If Hubble's constant is taken to be 70 km/sec/Mpc, and a quasar is found with a recessional velocity of 0.75 times the speed of light, how far away is the quasar?
   a. 3,200 Mpc
   b. 0.011 Mpc
   c. 52.5 Mpc
   d. 93 Mpc
   e. $1.6 \times 10^7$ Mpc

10. If a quasar has a red shift of 3, then the recessional velocity of the quasar is
   a. one-third the speed of light.
   b. three times the speed of light.
   c. 88 percent the speed of light.
   d. one-ninth the speed of light.
   e. one-fourth the speed of light.

**Fill in the Blank Questions**

11. Mid-sized black holes of a few thousand solar masses have been found in the cores of some _____ _____.

12. The central galaxy in a double-lobed radio source is usually a _____ galaxy.

13. Some _____ have fuzz around them that produces spectra similar to normal galaxies.

14. Double-lobed radio galaxies occur when jets of hot material interacts with the _____ _____.

15. In visible light images quasars resemble _____.

**True-False Questions**

16. Quasars must be small because they are surrounded by quasar fuzz.

17. A BL Lac object is observed if our line of sight is along the rotation axis of an active galactic nucleus.

18. The discovery of the gravitational lens effect for quasars shows that quasars are much further away than the distant galaxy that forms the gravitational lens.

19. Seyfert galaxies are three times more likely to be found as isolated galaxies as opposed to colliding galaxies.

20. The unified model for active galactic nuclei describes blazars, Type 1 Seyferts, Type 2 Seyferts, and radio lobe galaxies as being produced by a dense disk of gas and a hot accretion disk around a super massive black hole at the core of a galaxy.

## ADDITIONAL READING

Bechtold, Jill "Shadows of Creation: Quasar Absorption Lines and the Genesis of Galaxies." *Sky and Telescope 94* (Sept. 1997), p. 28.

Beck, Sara C. "Dwarf Galaxies and Starbursts." *Scientific American 283* (June 2000), p. 66.

Disney, Michael "A New Look at Quasars." *Scientific American 279* (June 1998), p. 52.

Garlick, Mark A. "Quasars Next Door." *Astronomy 29* (July 2001), p. 34

Shomaker, William "Going Deep for Galaxies." *Astronomy 29* (May 2001), p. 42.

Veilleux, Sylvain, Gerald Cecil, and Jonathan Bland-Hawthorn "Colossal Galactic Explosions." *Scientific American 277* (June 1996), p. 80.

Voit, Mark "The Rise and Fall of Quasars." *Sky and Telescope 97* (May 1999), p. 40.

## ANSWERS TO MORE MATH RELATED PROBLEMS

1. a. $0.581c = 174,000$ km/s
   b. 2,490 Mpc
2. a. 157.5 nm
   b. 0.24
   c. $0.21c = 6.3 \times 10^4$ km/s
   d. 900 Mpc
3. Fill in the table below, assuming that the Hubble constant is 70 km/s/Mpc.

| Object | Red shift | $V_r/c$ | $V_r$ (km/s) | Distance (Mpc) |
|--------|-----------|---------|--------------|----------------|
| 1 | 3.2 | 0.89 | $2.7 \times 10^5$ | 3,800 |
| 2 | 0.15 | 0.14 | $4.2 \times 10^4$ | 600 |
| 3 | 6.5 | 0.97 | $2.9 \times 10^5$ | 4,100 |
| 4 | 2.9 | 0.88 | $2.6 \times 10^5$ | 3,800 |
| 5 | 1.8 | 0.77 | $2.3 \times 10^5$ | 3,300 |
| 6 | 0.47 | 0.37 | $1.1 \times 10^5$ | 1,600 |

4. $0.016$ Mpc = 16 kpc
5. a. $1.2$ Mpc = $3.7 \times 10^{19}$ km
   b. $8.6 \times 10^{15}$ s = $2.7 \times 10^8$ years

**ANSWERS TO PRACTICE TEST**

1. c
2. d
3. e
4. d
5. b
6. b
7. b

8. e
9. a
10. c
11. globular clusters
12. peculiar or distorted
13. quasars
14. intergalactic medium

15. stars
16. F
17. T
18. T
19. F
20. T

# C H A P T E R   1 5

# COSMOLOGY IN THE 21ST CENTURY

## UNDERSTANDING THE CONCEPTS

**15-1  Introduction to the Universe**
  **The Edge-Center Problem**
    Why do modern cosmologist assume that the universe does not have an edge?
    What is the central concept of this sub-section?
  **The Necessity of a Beginning**
    What is Olbers' paradox?
    What does the solution to Olbers' paradox tell us about our universe?
    What is the difference between the universe and the observable universe?
    What is the central concept of this sub-section?
  **Window on Science 15-1.  Thinking About an Abstract Idea: Reasoning by Analogy**
    How can analogies help us in science?
    Why are analogies very important in our discussion of cosmology?
    What is the central concept of this sub-section?
  **Cosmic Expansion**
    What evidence tells us that the universe is expanding uniformly?
    What is the central concept of this sub-section?
  **The Necessity of a Big Bang**
    What is the big bang?
    Why is important NOT to think of the big bang as an explosion?
    What is meant by the Hubble time?
    What is the central concept of this sub-section?
  **Window on Science 15-2.  Why Scientists Speak Carefully: Words Lead Thoughts**
    Why must we be especially careful when discussing cosmology to chose our words carefully?
    What is the central concept of this sub-section?
  **The Cosmic Background Radiation**
    What is the cosmic microwave background radiation?
    What is the measured temperature of the cosmic microwave background radiation?
    At what temperature was this radiation produced?
    What causes the difference between the measured temperature and the temperature at which the radiation was produced?
    What is the central concept of this sub-section?
  **The Story of the Big Bang**
    What is meant when we say that photons have a temperature of 1 trillion K?
    How can we define the density of the universe at a time when it only contained photons?
    What do we mean when we say that the photons no longer had enough energy to produce protons or neutrons?
    During what time period were all the protons, neutrons, and electrons of the universe produced?
    Why were elements heavier than hydrogen and helium produced in the early universe?
    What produced the process produced the radiation we see today as the microwave background radiation?
    What is the central concept of this sub-section?

What are the fundamental ideas presented in this section?

**15-2 The Shape of Space and Time**
    **Looking at the Universe**
        What is isotropy?
        What is homogeneity?
        What is the cosmological principle?
        What is the central concept of this sub-section?
    **The Shape of the Expanding Universe and The Nature of Space-Time on pages 308 and 309**
        What causes the red shift we see in distant galaxies?
        How can a finite universe have no center?
        What is the central concept of this sub-section?
    **Model Universes**
        What is meant by the term "critical density"?
        What are the fates of an open universe, a flat universe, and a closed universe?
        If the universe is flat or closed, what problem does that pose for the age of the universe?
        What is the central concept of this sub-section?
    **Dark Matter in Cosmology**
        Why is dark matter a critical issue in determining the density of the universe?
        What is nonbaryonic matter?
        What is hot dark matter and cold dark matter?
        Do observations indicate that enough dark matter exists to make the density of the universe equal to the critical density?
        What is the central concept of this sub-section?

What are the fundamental ideas presented in this section?


**15-3 21$^{st}$ Century Cosmology**
    **Inflation**
        What is the flatness problem?
        What is the horizon problem?
        What are grand unified theories?
        What is suggested as the cause of the inflation in the inflationary universe theory?
        What does the inflationary theory predict about the density and geometry of the universe?
        What is the central concept of this sub-section?
    **The Acceleration of the Universe**
        What evidence indicates that the universe is expanding?
        What is the cosmological constant?
        What is quintessence?
        What is dark energy?
        What is the central concept of this sub-section?
    **The Origin of Structure and the Curvature of the Universe**
        What is a filament?
        Why are filaments a problem for cosmologists?
        What does the size of irregularities in the cosmic microwave background radiation tell us about the type of universe we live in?
        What is the central concept of this sub-section?

What are the fundamental ideas presented in this section?

## KEY CONCEPTS

This chapter discusses the large-scale structure of the universe and the theories that modern science has developed to explain the structure and evolution of the universe. This chapter relies heavily on the preceding three chapters. The material of this chapter requires extra preparation. The ideas presented are often at odds with common sense and this can lead to confusion in trying to understand concepts and conclusions.

There are two principle assumptions in modern cosmology; that the universe is homogeneous and isotropic. These two assumptions lead to a universe that has no edge and consequently no center. It is important to realize that while it seems natural to assume that the universe has an edge, the edge of the universe is unphysical.

Olbers' paradox shows us that a simple observation, the dark night sky, tells us something important about the structure of our universe; it is not infinitely old. Olbers' paradox also shows us the importance of identifying and testing the assumptions that a theory or idea are based upon.

The assumptions of homogeneity and isotropy are crucial to understanding the big bang and inflationary theories. It is important that we look for homogeneity and isotropy on large scales, not the small naked eye scales visible in the night sky. Homogeneity is seen in the Hubble expansion and isotropy is seen in the cosmic background radiation.

The second section looks at what we have observed in our universe and how different models might fit those observations. The section looks at the fundamentals of cosmology including homogeneity, isotropy, the critical density, the curvature of the universe, and the importance of dark matter. It provides the bases for our understanding of the origin of the universe and for the interpretation of the observed acceleration of our universe.

The final section presents the most recent findings in observational cosmology and their ramifications to our understanding of the origin and evolution of the universe. Inflation solves two long-standing problems, the horizon problem and the flatness problem. The discovery that our universe is currently in a state of acceleration leads to major questions. What could cause the acceleration? What does this imply about the age of the universe? Does this call into question our assumptions of homogeneity and isotropy, or does it support them? We are currently wrestling with these questions. The data suggests that there is a form of dark energy that is causing an acceleration of the expansion of the universe, but the form of that dark energy and its impact on the structure and future of the universe are very uncertain. This chapter, better than all others, demonstrates the dynamic nature of science and its continual search for understanding as new evidence is gleaned from nature.


## WEBSITES OF INTEREST

http://map.gsfc.nasa.gov/html/web_site.html          Microwave Anisotropy Probe
http://www.astro.ucla.edu/~wright/cosmolog.htm          Great Introduction to Cosmology
http://www.ncsa.uiuc.edu/Cyberia/Cosmos/CosmicMysteryTour.html          Great Cosmology Intro
http://www.damtp.cam.ac.uk/user/gr/public/cos_home.html          Cambridge Cosmology Site

### QUESTIONS ON CONCEPTS

1.  What evidence tells us that the universe began with a big bang?

    Solution:
    The primary evidence is threefold.  First the night sky is dark and not brilliantly bright.  This implies that the universe cannot be infinitely old; it has a beginning.  The dark night sky doesn't tell us anything about how the universe began, but it does imply that there was a beginning.

    The second piece of evidence suggesting a big bang is the correlation between the red shift of a galaxy and its distance.  At greater and greater distances, we see the universe at earlier and earlier times.  What we see is that the universe is expanding.  This expansion implies that there exists a time when all of the material was together with a very large density.

    The third piece of evidence is the cosmic microwave background radiation.  This radiation indicates that the universe is filled with radiation from the recombination of hydrogen, that is the formation of hydrogen atoms as electrons were bound in orbits around protons.  This indicates that the universe at one time was dominated by radiation.  The cosmic microwave background radiation is a signature of the changing nature and cooling of our universe.

    Any one of these pieces of information is very limited in what it tells us about the universe.  As the three are stitched together in a scientific argument, a clear case is made for the formation of the universe through a big bang.

2.  Why would a Hubble constant equaling 100 km/s/Mpc cause problems for most astronomers?

    Solution:
    The constant itself would not be a problem.  However, a Hubble constant of 100 km/s/Mpc implies a Hubble time of only 10 billion years.  The universe couldn't be older than 10 billion years, even if the universe was open, and a flat universe would be no more than 6.7 billion years old.  Our models of stellar evolution and observations of globular clusters tell us that most of the globular clusters in our own galaxy appear to be in excess of 10 billion years old, with some a little more than 13 billion years old.  It would be a real problem to have objects in the universe that are older than the universe itself.  A Hubble constant of 100 km/s/Mpc would be (and was) closely scrutinized by astronomers.

### WORKED EXAMPLES OF PROBLEMS REQUIRING MATHEMATICS

1.  If a galaxy is at a distance of 20 Mpc and has a recessional velocity of 1,300 km/s, what is the Hubble constant, H?

    Solution:
    We know
    the distance to the galaxy = $d$ = 20 Mpc and
    the recessional velocity = $V_r$ = 1,300 km/s.

    We can use the Hubble law in *By the Numbers 13-1* page 263 to find the Hubble constant.

    $V_r = H \cdot d$  If we divide both sides by $d$, we can solve for $H$.  $\dfrac{V_r}{d} = \dfrac{H \cdot d}{d} = H$

    Rearranging and entering our known values, we find

$$H = \frac{V_r}{d} = \frac{1{,}300 \text{ km/s}}{20 \text{ Mpc}} = 65 \frac{km/s}{Mpc}$$

Based on our data, the Hubble constant would be 65 km/s/Mpc.

2.   What is the density (in g/cm$^3$) of a cluster of galaxies that is 10 Mpc in radius that contains 1000 galaxies. Assume the average galaxy has a mass of $1 \times 10^{11}$ solar masses? (Hint: the volume of a sphere is $^4/_3 \pi r^3$, the mass of the sun is $2.0 \times 10^{33}$ g, and 1 pc = $3.1 \times 10^{18}$ cm.)

Solution:
We know
the number of galaxies = 1000,
the mass of one galaxy = $1.0 \times 10^{11}$ solar masses = $1.0 \times 10^{11} \cdot 2.0 \times 10^{33}$ g = $2.0 \times 10^{44}$ grams per galaxy,
the radius of the cluster = $r$ = 10 Mpc = $1.0 \times 10^7$ pc $\cdot$ $3.1 \times 10^{18}$ cm/pc = $3.1 \times 10^{25}$ cm.

The total mass is the number of galaxies times the mass of the average galaxy.

$$Mass = (1000 \text{ galaxies}) \cdot \left(2.0 \times 10^{44} \tfrac{g}{galaxy}\right) = 2.0 \times 10^{47} \text{ g}$$

We know the density of an object is the mass divided by the volume, and that the volume $= V = {}^4/_3 \pi r^3$. The volume of the cluster is

$$V = \tfrac{4}{3}\pi r^3 = \tfrac{4}{3}\pi \cdot \left(3.1 \times 10^{25} \text{ cm}\right)^3 = 1.25 \times 10^{77} \text{ cm}^3$$

The density is then

$$density = \frac{Mass}{Volume} = \frac{2.0 \times 10^{47} \text{ g}}{1.25 \times 10^{77} \text{ cm}^3} = 1.6 \times 10^{-30} \tfrac{g}{cm^3}$$

The density of the cluster is $1.6 \times 10^{-30}$ g/cm$^3$.

3.   A few years ago, data suggested that the Hubble constant might be as high as 100 km/s/Mpc. What is the maximum age of the universe if $H = 100$ km/s/Mpc?

Solution:
We know the Hubble constant = $H$ = 100 km/s/Mpc. On page 300 there is a relationship for the maximum age of the universe as a function of the Hubble constant. The maximum age of the universe = the Hubble time = $T$, which is given by

$$T \approx \frac{1}{H} \times 10^{12} \text{ years} = \frac{1}{100} \times 10^{12} \text{ years} = 1 \times 10^{10} \text{ years}$$

If the Hubble constant is 100 km/s/Mpc, then the maximum age of the universe is only $1 \times 10^{10}$ years, which is notably younger than many globular clusters.

# MORE MATH RELATED PROBLEMS
(Answers at the end of the chapter)

1.  The distances and recessional velocities are measured for six galaxies. Find the Hubble constant for each galaxy and the average Hubble constant for the group.

| Object | $V_r$ (km/s) | $d$ (Mpc) | $H$ (km/s/Mpc) |
|--------|--------------|-----------|----------------|
| A | 1,625 | 25 | |
| B | 17,520 | 240 | |
| C | 1,309 | 17 | |
| D | 414 | 6 | |
| E | 57,600 | 800 | |
| F | 42,900 | 650 | |
| | | Average | |

2.  a.  What is the Hubble time if the Hubble constant is 50 km/s/Mpc?
    b.  What is the age of a flat universe if the Hubble constant is 50 km/s/Mpc?

3.  What is the density of a spherical galaxy in g/cm$^3$ if it contains $5 \times 10^{11}$ solar masses of material and has a radius of 12 kpc?
    Assume the galaxy is a perfect sphere. The volume of a sphere is given by $^4/_3 \pi r^2$.

4.  What is the density of a cluster of galaxies if it contains 300 galaxies, has a radius of 2.5 Mpc, and the average galaxy has a mass of $1.0 \times 10^{11}$ solar masses.

# PRACTICE TEST
## Multiple Choice Questions

1.  The cosmological principle states that
    I.   the universe is flat.
    II.  the universe is isotropic.
    III. the universe is accelerating.
    IV.  the universe is homogeneous.

    a.  I & II
    b.  I & IV
    c.  II & III
    d.  II & IV
    e.  III & IV

2.  If the density of the universe is less than the critical density, the universe is
    a.  open with positive curvature.
    b.  open with negative curvature.
    c.  closed with positive curvature.
    d.  closed with negative curvature.
    e.  flat with zero curvature.

3.  The _____ theories are theories that unify the electromagnetic, weak and strong forces at extremely high energies.
    a.  steady-state
    b.  inflationary
    c.  Newtonian
    d.  general relativity
    e.  grand unified

4.  The red shifts of the galaxies imply that the universe is
    a.  closed.
    b.  flat.
    c.  homogeneous.
    d.  isotropic.
    e.  expanding.

5.  That the density of the universe is very close to the critical density of the universe, is know as
    a.  the horizon problem.
    b.  Olbers' paradox.
    c.  the cosmological principle.
    d.  the flatness problem.
    e.  the oscillating universe theory.

6.  _____ can combine with a proton and annihilate itself and the proton to form a gamma-ray.
    a.  An electron
    b.  An antimatter proton
    c.  A positron
    d.  An antimatter neutrino
    e.  A neutrino

7.  During the early history of the universe _____ dominated the universe.
    a.  matter
    b.  photons
    c.  recombination
    d.  antimatter
    e.  cold dark matter

8.  The Hubble time is
    a.  the time from the initial expansion of the universe until the end of the inflationary period.
    b.  the time for the universe to increase its size by a factor of 2.
    c.  equal to $10^{12}$ divided by the Hubble constant.
    d.  an estimate of the time it takes recombination to occur.
    e.  an estimate of the time it takes the universe to reach the critical density.

9.  If the Hubble constant, H, is found to be smaller at large distances than it is nearby, what does this imply about the universe?
    a.  The universe must be older than we suspect.
    b.  Matter in the universe is very important to its motion.
    c.  The expansion of the universe is accelerating.
    d.  all of the above
    e.  none of the above

10. If a galaxy is located at a distance of 57 Mpc and it is found to have recessional velocity of 4,390 km/sec, what is Hubble's constant based only on this galaxy?
    a.   77 km/sec/Mpc
    b.   57 km/sec/Mpc
    c.   4,447 km/sec/Mpc
    d.   4,333 km/sec/Mpc
    e.   0.013 km/sec/Mpc

**Fill in the Blank Questions**

11. If the universe is _____, then its age will be less than two-thirds of $1/H$.

12. _____is the name of the force suspected to cause the current acceleration in the rate of expansion of the universe.

13. What was the temperature of the universe when the recombination took place that produced the cosmic background radiation we observe today?

14. The assumption of _____states that the universe looks the same from all locations over sufficiently great distances.

15. Current observations of the amount of baryonic matter and dark matter in the universe imply that the universe is _____.

**True-False Questions**

16. The resolution of Olbers' paradox suggests that it gets dark at night because the universe is closed.

17. Galaxy seeds, around which galaxies, clusters and walls grew, may have been caused by the separation of the electromagnetic and weak forces.

18. The cosmic background radiation was produced during the inflationary period.

19. If galaxy A is four times more distant than galaxy B, then according to the Hubble Law, galaxy A will recede 4 times slower than galaxy B.

20. During the first moments of the big bang, nuclear fusion reactions made few elements heavier than helium because no stable nuclei exist with masses of 5 or 8 hydrogen masses.

**ADDITIONAL READING**

Bennet, Charles L., Gary F. Hinshaw, and Lyman Page "A Cosmic Cartographer." *Scientific American 284* (Jan. 2001), p. 44.

Caldwell, Robert R. and Marc Kamionkowski "Echos from the Big Bang." *Scientific American 284* (Jan. 2001), p. 38.

Hawking, Stephen W. and Roger Penrose "The Nature of Space and Time." *Scientific American 277* (July 1996), p. 60.

Magueijo, João "Plan B for the Cosmos." *Scientific American 284* (Jan. 2001), p. 59.

Ostriker, Jeremiah P. and Paul J. Steinhardt "The Quintessential Universe." *Scientific American 284* (Jan 2001), p. 46.

Peebles, P. James E. "Making Sense of Modern Cosmology." *Scientific American 284* (Jan 2001), p. 54.

**ANSWERS TO MORE MATH RELATED PROBLEMS**

1. Complete the table

| Object | $V_r$ (km/s) | $d$ (Mpc) | $H$ (km/s/Mpc) |
|--------|--------------|-----------|----------------|
| A | 1,625 | 25 | 65 |
| B | 17,520 | 240 | 73 |
| C | 1,309 | 17 | 77 |
| D | 414 | 6 | 69 |
| E | 57,600 | 800 | 72 |
| F | 42,900 | 650 | 66 |
| | | Average | 70.3 |

2. a. 20 billion years
   b. 13 billion years
3. $4.6 \times 10^{-24}$ g/cm$^3$
4. $3.1 \times 10^{-29}$ g/cm$^3$

**ANSWERS TO PRACTICE TEST**

1. d
2. b
3. e
4. e
5. d
6. b
7. b
8. c
9. c
10. a
11. closed
12. Quintessence
13. 3,000 K
14. homogeneity
15. open
16. F
17. F
18. F
19. F
20. T

# C H A P T E R  1 6

# THE ORIGIN OF THE SOLAR SYSTEM

## UNDERSTANDING CONCEPTS

**16-1  The Great Chain of Origins**

    **A Review of the Origin of Matter**

        Where did your thumb come from?

        What is the central concept of this sub-section?

    **The Origin of Planets**

        What evidence do we have that disks of gas and dust surround stars before they reach the main sequence?

        What is the solar nebula theory?

        What is the central concept of this sub-section?

    **Window on Science 16-1.  Two Kinds of Theories: Evolution and Catastrophe**

        What is an evolutionary theory?

        What is a catastrophic theory?

        What is the central concept of this sub-section?

    **Planets Orbiting Other Stars**

        What are extrasolar planets?

        Why is finding extrasolar planets and dust disk around young stars important?

        In what ways do the extrasolar planets discovered to date, seem odd compared with the planets of the solar system?

        What is the central concept of this sub-section?

    **Window on Science 16-2.  Scientists: Courteous Skeptics**

        Why is it necessary of scientists to be skeptical?

        What is the central concept of this sub-section?

What are the fundamental ideas presented in this section?

**16-2  A Survey of the Solar System**

    **Revolution and Rotation**

        What is the difference between rotation and revolution?

        What is the central concept of this sub-section?

    **Two Kinds of Planets and Terrestrial and Jovian Planets pages 332 and 333**

        What are the major differences between the terrestrial and Jovian planets?

        Are craters found throughout the solar system or only on objects near the sun?

        What is the central concept of this sub-section?

    **Space Debris**

        What are asteroids?

        Where did they come from?

        How big is the nucleus of a comet?

        What is the central concept of this sub-section?

    **The Age of the Solar System**

        What is the half-life of a radioactive element?

        Based on radioactive ages, what is the age of the oldest rock found on Earth?

        Based on evidence from rocks and meteorites, what is the approximate age of the solar system?

        What is the central concept of this sub-section?

What are the fundamental ideas presented in this section?

### 16-3 The Story of Planet Building

**Window on Science 16-3. Reconstructing the Past Through Science**

What is the central concept of this sub-section?

**The Chemical Composition of the Solar Nebula**

What percentage of the mass of the sun is hydrogen?

What does the term gravitational collapse mean in the context of planet formation?

What is the central concept of this sub-section?

**The Condensation of Solids**

What is the uncompressed density of a planet?

What is the condensation sequence?

What does the condensation sequence tell us about the formation of the terrestrial planets?

What is the central concept of this sub-section?

**The Formation of Planetesimals**

What is condensation?

What is accretion?

Why are planetesimals thought to collapse to the plane of the forming solar system, but not individual atoms and molecules?

What is the central concept of this sub-section?

**The Growth of Protoplanets**

Why would large planetesimals grow faster than smaller ones?

What is differentiation and under what condition can it occur?

What is outgassing?

What is the heat of formation and what role might it play in the growth of planets?

Why do we believe that Jovian planets grew to their present size in about 10 million years?

What is the central concept of this sub-section?

**Explaining the Characteristics of the Solar System**

Why does the solar system have a disk shape?

How does the condensation sequence explain the differences between the terrestrial and Jovian planets?

Why were the Jovian planets able to capture atmospheres from the solar nebula, but the terrestrials were not able to do so?

Why are there asteroids between Mars and Jupiter and not a planet?

What is the central concept of this sub-section?

**Clearing the Nebula**

What is radiation pressure?

How did the solar wind help clear material from the solar nebula?

What evidence supports the sweeping up of material from the solar nebula by the planets?

What are four methods for clearing the solar nebula?

What is the central concept of this sub-section?

What are the fundamental ideas presented in this section?

## KEY CONCEPTS

There are two major topics in this chapter that are presented in a tightly correlated manner. The first is the solar nebula theory, which is the theory that currently best describes the formation of the solar system in light of the available data. It is important to realize that the solar system has a structure that can tell us something about how it formed. Therefore, you need to understand the differences between terrestrial planets, Jovian planets, and the minor objects of the solar system. The physical concepts of a condensation sequence, accretion, and differentiation are important in understanding how the solar nebula theory describes the formation of the solar system. These are the major processes in planet building.

The second topic in this chapter is an overview of the solar system, which provides an introduction to the solar system and the data necessary to analyze the solar nebula theory. The solar system shows a great deal of order, from the division of planets into terrestrials and Jovians, to the consistent revolution directions of the planets. This order allows us to analyze the data and construct a theory for the formation of the solar system. It is from the observation of basic physical properties and the classification of the objects of the solar system into groups that we are able to understand the processes that formed the solar system.

One final point, the formation of planets appears at present to be best described by an evolutionary theory, as opposed to a catastrophic theory. Therefore, planets and planetary systems should be common around other stars throughout the galaxy. The fact that we have begun to find planets around nearby stars supports an evolutionary theory for the formation of the solar system.


## WEBSITES OF INTEREST

http://seds.lpl.arizona.edu/nineplanets/nineplanets/nineplanets.html     Tour of the Solar System
http://exoplanets.org/     Extra Solar Planets
http://www.jpl.nasa.gov/     Good NASA Planetary site


## QUESTIONS ON CONCEPTS

1.  What do the condensation sequence and the uncompressed density of solar system objects tell us about the formation of the solar system?

    Solution:
    The condensation sequence is simply a list of the temperatures at which various materials condense, that is, change phase from gas to solid. The existence of certain types of material in a planet indicates that the temperature of the solar nebula at that location was less than the condensation temperature of those materials at the time the nebula was cleared.

    The uncompressed density of the planets provides information about the types of materials in the planets as a function of distance from the sun. We find that the uncompressed density of the planets decreases as the distance to a planet increases. This implies that there were few ices in the inner solar system at the time the solar nebula was cleared, and that ices and other volatile materials were still in the solar nebula at the location of Jupiter's orbit (about 5 AU).

    We would expect that the solar nebula was hottest near the sun and cooled as the distance from the sun increased. These all fit nicely together adding evidence to the solar nebula theory. We expect the temperature to decrease at greater distances from the sun, and we expect certain types of materials to be present depending on the temperature of the material at the time the nebula was cleared. The uncompressed densities of the planets support this, and indicate the approximate temperature of the solar nebula at the time it was cleared.

2.  Why are materials with half-lives greater than 1 billion years used to determine the age of solar system rock samples?

    Solution:
    Most of our samples are a few billion years old. We want to use radioactive materials to date the samples that have half-lives approximately equal to the age of the material we are dating. If the half life is very short, most of the material will have decayed and it will be difficult to determine the exact abundance. If the half-life is very long, then a tiny portion will have decayed. As an example, a radioactive material that has a short half-life of say 10 million years that originally contained $10^{30}$ atoms would only contain 1 atom after 1 billion years. Note that $10^{30}$ carbon atoms would have a weight of 22 tons. That's a pretty large sample.

## WORKED EXAMPLES OF PROBLEMS REQUIRING MATHEMATICS

1.  The diagram to the right illustrates the radioactive decay of Potassium ($^{40}K$), which has a half-life of 1.3 billion years. If a lunar rock is found that currently contains 0.08 grams of $^{40}K$, and it is determined that the sample contained 0.64 grams when it was formed, how old is the lunar rock?

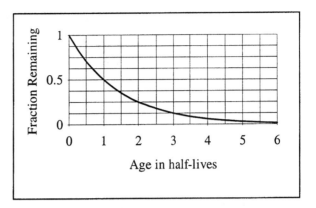

Solution:
We know
the half-life of $^{40}K = t_{1/2} = 1.3$ billion years,
the original mass of $^{40}K$ in the sample $= m_o = 0.64$ grams, and
the observed mass of $^{40}K$ in the sample $= m = 0.08$ grams.
The fraction of the original material remaining equals the current mass divided by the original mass.

$$\textit{fraction remaining} = \frac{\textit{current mass}}{\textit{original mass}} = \frac{m}{m_o} = \frac{0.08\,g}{0.64\,g} = \frac{1}{8} = 0.125.$$

Notice that the vertical axis of the graph is divided into eighths. The first horizontal line above the axis is at a fraction of $1/8^{th}$ of the original mass. If we follow this horizontal line over to where it meets the plotted curve, we find that the sample has an age of 3 half-lives. The age of the sample equals the number of half-lives times the time for one half-life.

$$\textit{age} = (\textit{number of half-lives}) \cdot (\textit{length of one half-life}) = 3 \cdot (1.3\text{ billion years}) = 3.9\text{ billion years}.$$

The rock would have an age of 3.9 billion years.

2.  Suppose that through accretion Mars grew to its present size in 10 million years, and that the mass of the average particle accreted was 0.10 kg. Mars' current mass is $6.4 \times 10^{23}$ kg.
    a.  What was the average accretion rate? That is, how much mass did Mars gain each hour?
    b.  On average, how many particles were captured each hour?

Solution to part a:
We know
the current mass of Mars $= M = 6.4 \times 10^{23}$ kg and
the time period for the accretion $= 10 \times 10^9$ years.
We convert to hours time $= (10 \times 10^9 \text{ years}) \cdot (365.25 \text{ days/year}) \cdot (24 \text{ hours/day}) = 8.8 \times 10^{13}$ hours.

The mass accreted in one hour = the current mass divided by the time it took to accrete that mass.

$$\textit{mass accreted in one hour} = \frac{\textit{current mass of Mars}}{\textit{time period for accretion}} = \frac{6.4 \times 10^{23}\text{ kg}}{8.8 \times 10^{13}\text{ hours}} = 7.3 \times 10^9 \text{ kg/hour}$$

The accretion rate is $7.3 \times 10^9$ kg/hour.

Solution to part b:
We know
the mass accreted in one hour = $7.3 \times 10^9$ kg/hour and
the mass of each particle accreted = $m$ = 0.10 kg.
We want to find the number of particles captured in one hour.
The number of particle captured in one hour = $n$ = the mass captured in one hour divided by the mass of one particle. Therefore

$$number\ of\ particles\ per\ hour = \frac{mass\ accreted\ in\ one\ hour}{mass\ of\ each\ particle} = \frac{7.3 \times 10^9 \frac{kg}{hr}}{0.10 \frac{kg}{particle}} = 7.3 \times 10^{10}\ particles/hour$$

About $7.3 \times 10^{10}$ particles hit Mars each hour under these assumptions.

3. A star, 8 pc from Earth, is found to have a planet located 10 AU from the star. What is the angular separation of the star and planet as viewed from Earth?

Solution:
We know
the distance to the star/planet system = 8 pc = 8 pc · 206,265 AU/pc = $1.65 \times 10^6$ AU and
the distance between the star and planet = 10 AU.

We use the small angle formula from *By the Numbers 3-1* on page 32 to find the angular separation

$$\frac{angular\ separation}{206,265"} = \frac{linear\ separation}{distance}$$

Multiplying both sides by 206,265" we can solve for the angular separation.

$$angular\ separation = 206,265" \cdot \left( \frac{linear\ separation}{distance} \right) = 206,265" \cdot \left( \frac{10\ AU}{1.65 \times 10^6\ AU} \right) = 1.25"$$

The angular separation between the star and planet would be 1.25 seconds of arc, which is measurable, but the large difference in brightness would make this extremely difficult.


**MORE MATH RELATED PROBLEMS**
(Answers at the end of the chapter)

1. If a sample contains 40 atoms of $^{40}$K and originally contained 160 atoms, what is the age of the sample? The half-life of $^{40}$K is 1.3 billion years.

2. If a rock contains 2 grams of a radioactive material and it can be determined that the sample originally contained 16 grams of the material, how many half-lives has it been since the rock formed?

3. The radioactive decay of a material is tested in a laboratory. It is found that the sample contains one-fourth of its original mass after 10 years. What is the half-life of this material?

4. Mercury's mass is $3.3 \times 10^{23}$ kg. If the hourly in-fall rate was $3.2 \times 10^{12}$ kg/hour, how long did it take Mercury to reach its present mass?

5. On a photograph of the moon, the moon measures 46 cm in diameter and a small crater measures 0.66 cm in diameter. The moon's physical diameter is 1738 km, what is the physical diameter of the small crater?

6. The solar wind travels at about 400 km/s. How long does it take particles in the solar wind to travel 20 AU? (1 AU = $1.5 \times 10^8$ km.)

**PRACTICE TEST**
**Multiple Choice Questions**

1. The original cloud from which the sun and planets formed had a composition
   a. similar to that of the asteroids.
   b. similar to that of the terrestrial planets.
   c. rich in elements heavier than helium.
   d. dominated by hydrogen and helium.
   e. dominated by oxygen and carbon.

2. The characteristics of the Jovian planets include
   I. low average density.
   II. orbits inside the asteroids.
   III. craters in old surfaces.
   IV  very few moons.

   a. I
   b. II
   c. II & IV
   d. I, II, & III
   e. II, III, & IV

3. The adding of material an atom at a time is the process of _____.
   a. accretion
   b. morphing
   c. condensation
   d. outgassing
   e. heavy bombardment

4. Which solar system objects have been dated using radioactive decay?
   a. Earth and the moon
   b. Mercury, Venus, and Earth
   c. The moon and several meteorites
   d. Jupiter, Earth and the moon
   e. Mars, Earth, the moon, and several meteorites

5. When a rocky or icy surface of a solar system object is found to have only a few craters, then astronomers know
   a. that the surface is young.
   b. that the surface is old.
   c. that the surface formed when there were very few asteroids.
   d. that the object has a very cold interior.
   e. that the object does not have an atmosphere.

6. The half-life of a radioactive material is equal to
   a. the time it takes the material to form.
   b. half the time it takes the material to form.
   c. the time it takes the nuclei in the material to completely decay.
   d. the time it takes half the nuclei in the material to decay.
   e. the age of the material.

7. The solar nebula was cleared of gas and dust and planet formation stopped when
   a. Jupiter became massive enough to begin to sweep hydrogen and helium into its atmosphere.
   b. water first appeared on Earth's system.
   c. the sun became a luminous object.
   d. a star passed near by and pulled much of the material out of the nebula.
   e. terrestrial planets began to accrete metal oxides and metals.

8. One of the mechanisms that helped clear the solar nebula following the formation of the planets was
   a. outgassing.
   b. differentiation.
   c. accretion.
   d. condensation
   e. the solar wind.

9. The speed of the solar wind is approximately 400 km/s and the distance from the sun to Earth is $1.5 \times 10^8$ km. How long does it take a particle in the solar wind to reach Earth?
   a. about 4.3 hours
   b. about 4.3 days
   c. about 43 days
   d. about 4.3 years
   e. about 43 years

10. The diagram to the right illustrates the radioactive decay of Potassium ($^{40}$K), which has a half-life of 1.3 billion years. If a lunar rock is found that currently contains 16 grams of $^{40}$K, and it is determined that the sample contained 32 grams when it was formed, how old is the lunar rock?
    a. 0.65 billion years
    b. 1.3 billion years
    c. 2.6 billion years
    d. 20.8 billion years
    e. 41.6 billion years

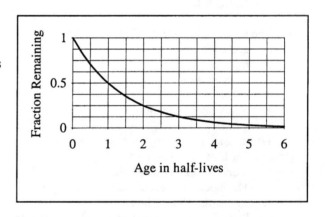

**Fill in the Blank Questions**

11. What solar system objects are most like the planetesimals that formed in the solar nebular?

12. A _____ is a solar system object that enters Earth's atmosphere and becomes very hot due to friction between the object and Earth's atmosphere.

13. _____ _____ is caused as light from the sun pushes on particles. This process helped clear the early solar nebula.

14. _____ refers to the formation of an atmosphere by the release of gases from the rocks and minerals of the planet.

15. A(n) _____ theory is based on slow processes that do not depend on unlikely events.

**True-False Questions**

16. The oldest rocks found on Earth are about 3.9 billion years old.

17. The uncompressed density of a planet in our solar system is greatest for the planets closest to the sun.

18. The condensation sequence suggests that ices of water, methane, and ammonia should condense closest to the sun.

19. Protoplanets of the Jovian planets could have grown very hot from radioactivity of light elements such as hydrogen and helium.

20. The current atmosphere of Earth is believed to be the remnants of the original gas from the solar nebula attracted by the protoplanet.

**ADDITIONAL READING**

Comins, Neil F. "We Are All Star Stuff." *Astronomy 29* (Jan. 2001), p. 56.

Doyle, Laurance R., Hans-Jörg Deeg, and Timothy M. Brown "Searching the Shadows of Other Earth's." *Scientific American 283* (Sept. 2000), p. 58.

Roger, J., P. Angel, and Neville J. Woolf "Searching for Life on Other Planets." *Scientific American 277* (April 1996), p. 60.

Wood, John A. "Forging the Planets: The Origin of Our Solar System." *Sky and Telescope 97* (Jan. 1999), p. 36.

**ANSWERS TO MORE MATH RELATED PROBLEMS**
1. 2.6 billion years
2. 3 half-lives
3. 5 years
4. $1.03 \times 10^{11}$ hours = 12 million years
5. 25 km
6. $7.5 \times 10^{6}$ s = 2,100 hours = 87 days

**ANSWERS TO PRACTICE TEST**

| | | |
|---|---|---|
| 1. d | 8. e | 15. evolutionary |
| 2. a | 9. b | 16. T |
| 3. c | 10. b | 17. T |
| 4. e | 11. asteroids | 18. F |
| 5. a | 12. meteor | 19. F |
| 6. d | 13. Radiation pressure | 20. F |
| 7. c | 14. Outgassing | |

## CHAPTER 17

# THE EARTHLIKE PLANETS

**UNDERSTANDING THE CONCEPTS**

**17-1  The Early History of Earth**
   **Four Stages of Planetary Development**
      What are the four major stages of planet formation that Earth passed through?
      What is meant by flooding?
      What is the central concept of this sub-section?
   **Earth's Interior**
      Earth's interior is divided into three general regions; describe each.
      What is the temperature and density of Earth's core?
      How do we know that Earth has a liquid core?
      What is the central concept of this sub-section?
   **Earth's Active Crust and The Active Earth on pages 350 and 351**
      What is plate tectonics?
      What causes most of the geologic activity we see on Earth?
      How can astronomers measure continental drift?
      What is the central concept of this sub-section?
   **Earth's Atmosphere**
      What was Earth's primeval atmosphere like and how did it form?
      What happened to the carbon dioxide originally in Earth's atmosphere?
      What is the greenhouse effect?
      What is the central concept of this sub-section?
   **Oxygen in Earth's Atmosphere**
      Why is photosynthesis important to Earth's atmosphere?
      What layer in Earth's atmosphere protects animal life from harmful ultraviolet radiation?
      What is the central concept of this sub-section?

   What are the fundamental ideas presented in this section?

**17-2  The Moon**
   **Lunar Geology and Impact Cratering on pages 354 and 355**
      What are maria?
      What are the distinguishing characteristics of impact craters?
      Why are there so many more impact craters on the moon than on Earth?
      What are breccias?
      What is the central concept of this sub-section?
   **The History of Earth's Moon**
      In what ways has the moon's small size impacted its history?
      What material flooded the lunar basins?
      What causes the erosion of the lunar surface?
      What is the central concept of this sub-section?
   **Window on Science 17-1.  Understanding Planets: Follow the Energy**
      Why is the flow of energy so important in astronomy?
      What is the central concept of this sub-section?

**The Origin of Earth's Moon**

   Which hypothesis for the formation of the moon is best supported by the observational evidence?

   How does the large-impact hypothesis describe the lack of an iron core in the moon?

   What is the central concept of this sub-section?

What are the fundamental ideas presented in this section?


## 17-3  Mercury

**Mariner 10 at Mercury**

   How is Mercury's surface like that of the moon?

   How is Mercury's surface not like that of the moon?

   How is Mercury's interior different from the moon's interior?

   What is the central concept of this sub-section?

**Window on Science 17-2.  How Hypotheses and Theories Unify the Details**

   How do the stories that scientists tell differ from those of works of fiction?

   What is the central concept of this sub-section?

**A History of Mercury**

   What effect has Mercury's small size had on its evolution?

   What evidence do we have that Mercury has a large iron core?

   How did the lobate scarps form?

   What is the central concept of this sub-section?

What are the fundamental ideas presented in this section?


## 17-4  Venus

**The Atmosphere of Venus**

   What are the major constituents of Venus' atmosphere?

   How does the density of Venus' atmosphere compare to Earth's?

   Why is the Venetian atmosphere so hot?

   What is the central concept of this sub-section?

**The Surface of Venus and Volcanoes on pages 366 and 367**

   What are the two main types of volcanoes?

   What type of volcanoes are found on Venus?

   What does the type of volcanoes found on Venus tell us about the planet?

   Based on the craters found on Venus, what is the age of the Venetian surface?

   What is the central concept of this sub-section?

**The History of Venus**

   What is the primary cause of the differences between Earth's and Venus' atmospheres?

   What affect might the high temperature on Venus have had on the geology of Venus?

   What is the central concept of this sub-section?

What are the fundamental ideas presented in this section?


## 17-5  Mars

**The Atmosphere of Mars**

   What is the composition and density of Mars' atmosphere in comparison to Earth's and Venus'?

   Why can't liquid water exist on the Martian surface?

   Why is the Martian atmosphere so thin?

   What is the central concept of this sub-section?

**The Geology of Mars**

   What type of volcano is found on Mars?

   What does Olympus Mons tell us about the crust of Mars?

   What is the central concept of this sub-section?

**Hidden Water on Mars**
    What are outflow channels and runoff channels?
    What does the deuterium on Mars tell us about water on Mars?
    What is the central concept of this sub-section?
**The History of Mars**
    What is the core of Mars believed to be like?
    What is the central concept of this sub-section?
**Window on Science 17-3. The Present is the Key to the Past**
    What is the final goal of planetary astronomy?
    What is the central concept of this sub-section?
**The Moons of Mars**
    What are the surfaces of Deimos and Phobos like?
    What is the central concept of this sub-section?

What are the fundamental ideas presented in this section?


## KEY CONCEPTS

Earth is studied as part of an astronomy course for three reasons; it is our home planet, we know more about it than any other planet, and what we learn about Earth's structure, internal dynamics, and atmosphere are applicable to our study of the other planets, especially the terrestrials. Therefore, this chapter and our study of the bodies of the solar system begin with a study of planet Earth.

There are four basic stages in the development of a terrestrial planet following its formation. These are differentiation, cratering, flooding, and slow surface evolution. All of the terrestrial planets went through these four stages, though to different extents.

All of the terrestrial planets differentiated to form an iron core, a mantle and a rocky crust. The evidence for this comes primarily from their mean densities, which are all about 5 g/cm$^3$.

Cratering is observed on all of the terrestrial planets and the moon and provides us with a way to determine the age of planetary surfaces. Surfaces that contain several large impacts and numerous small impacts, like the surfaces of the moon and Mercury, tell us that theses surfaces formed and remain largely unchanged since the time of heavy bombardment 4 billion years ago. The cratering record on Mars indicates that the southern highlands are much older than the northern plains, suggesting that some processes erased many craters in the northern plains region. The cratering record on Venus tells us that the age of the Venetian surface is uniform and fairly young. The uniform cratering record implies that some process completely resurfaced Venus in the last half-billion years.

Flooding can be due to molten material or water or any other semi-liquid substance. In the case of Earth most of the flooding continues and is due to water. Some flooding by lava is seen, though it is minor. On the moon the maria show huge regions that were flooded by basaltic lava flows following crater impacts. Similar flooding dominated Mercury as well. On Venus we find a much younger surface than on the moon or Mercury, but we still see large basaltic plains indicating that they were flooded by lava flows. These are most likely the result of volcanism since large impact craters do not mark the surface. Additionally, Venus has several large volcanic mountains that implicate volcanism as a cause for the flooding by lava that we observe. Mars shows some evidence of flooding of impact craters by lava, flooding by volcanic eruptions, and flooding by water.

Slow surface evolution is limited to impacts by small particles on Mercury and the moon. Without atmospheres or active crusts, Mercury and the moon have changed little in the last 4 billion years, except through impacts. Venus, Mars, and Earth show signs of active geology, but only Earth shows plate tectonics. Mars and Earth both show evidence of slow changes due to flowing water, i.e. water erosion. Wind erosion is expected to play a small role in slow surface evolution on Venus, Earth, and Mars.

The atmospheres of the terrestrial planets show some significant and interesting variations. Mercury is too hot and too small to retain an atmosphere. The compositions of the atmospheres of Venus and Mars are similar, yet Earth's is much different. The appearance of liquid water early in Earth's history greatly influenced its future atmosphere. This water allowed much of the carbon dioxide to be dissolved from the atmosphere, leaving Earth with an atmosphere dominated by nitrogen. The reduced carbon dioxide abundance kept the greenhouse effect from pushing temperatures to the levels seen on Venus.

It is important to understand that the surface features, composition, and atmosphere of each planet tell us about the planet, and provide evidence to build theories about the planet's evolution and the formation of the solar system as a whole. Remember that one of the prime goals of astronomy is to better understand nature so that we can apply what we learn about nature in our "backyard" to make predictions and to better understand more distant and complex objects.

## WEBSITES OF INTEREST

| | |
|---|---|
| http://pao.gsfc.nasa.gov/gsfc/earth/earth.htm | NASA missions to observe Earth |
| http://www.jpl.nasa.gov/solar_system/ | NASAs JPL planetary site. |
| http://seds.lpl.arizona.edu/nineplanets/nineplanets/ | Planets Site |

## QUESTIONS ON CONCEPTS

1. Why are there so many more impact craters on the moon than there are on Earth or Mars?

Solution:
The biggest reason for the large number of craters on the moon and the very small number of craters on Earth and Mars is geologic activity. The moon is geologically inactive and has been for several billion years. Craters that formed during the period of heavy bombardment 4 billion years ago are still visible. The only process altering the surface of the moon over the last 4 billion years is the impact of material. Particles larger than a few cm create craters and smaller particle impacts slowly erode the surface features.

In contrast the surface of Mars has been active during the past. The surface of Mars shows clear signs of volcanism with lava flows. The surface also suggests that water was once present and may have flooded a large portion of the planet within the last 2 billion years.

The surface of Earth has very few visible craters. The surface of Earth is very active. Volcanism, plate tectonics, wind erosion, and water erosion are all continually changing the appearance of Earth's surface. Consequently, old impact craters have been erased.

It is tempting to point to the atmospheres of Earth and Mars for the difference in the number of craters between them and the moon, but the atmospheres play a very limited role. Looking more closely at the surface of Mars supports this. The southern highlands on Mars show heavy cratering and a surface more like the moon's. The cratering record suggests that the southern highlands are between 2 and 3 billion years old. Clearly, the atmosphere of Mars has not protected this region from impacts and crater formation. Therefore the young surface found in the northern lowland on Mars contains few craters because of processes that actively altered the surface and erased past impact craters.

2. Why did the atmosphere of Venus develop differently from that of Earth?

Solution:
The biggest difference between Earth's atmosphere and Venus' is the level of carbon dioxide. The early atmospheres of these two planets is believed to have been very similar. Earth contained water vapor in its atmosphere which condensed to form oceans when the temperature fell below the boiling point. Carbon dioxide dissolves easily in water and combines with other minerals to form limestone and other mineral deposits. This process removed a great deal of the carbon dioxide from Earth's atmosphere. Since Venus did not cool enough to allow water to condense, the carbon dioxide remained in its atmosphere. This higher level of carbon dioxide traps a great deal more heat and has elevated the temperature on Venus.

3. What factors determine the rate of loss of gases from a planet's atmosphere?

Solution:
Three factors determine the ability of a planet to retain certain elements within its atmosphere. These factors are the escape velocity of the planet, the mass of the atom/molecule in the atmosphere, and the temperature of the planet's atmosphere. The escape velocity depends on the mass and radius of the planet. If the planet has a large mass it will have a fairly strong gravitational field and will be able to retain a thick atmosphere of relatively light particles. The gravitational force of attraction also depends on the mass of the particle in the atmosphere. The force is greater on massive atoms/molecules, so it is easiest to retain heavy molecules. Finally, the temperature of the planet determines the speed at which particles move in the atmosphere. The higher the temperature the faster the particles move. Additionally, the massive atoms/molecules will move more slowly than the lighter ones. These factors can be seen in Figure 17-15.

## WORKED EXAMPLES OF PROBLEMS REQUIRING MATHEMATICS

1. The smallest detail observable with Earth-based telescopes is about 1 second of arc in diameter? What size does this correspond to on the surface of Mars at its closest approach to Earth? (Hint. At its closest approach to Earth, Mars is at a distance of 0.52 AU.)

Solution:
We know
the angular diameter = 1 second of arc and
the distance = 0.52 AU = (0.52 AU) $\cdot$ (1.5$\times10^8$ km/AU) = 7.8$\times10^7$ km.

We can use the small angle formula from *By the Numbers 3-1* on page 32 to find the linear size of a feature.

$$\frac{linear\ diameter}{distance} = \frac{angular\ diameter}{206,265"}$$

If we multiply both sides by the distance, we can solve for the linear diameter.

$$linear\ diameter = distance \cdot \frac{linear\ diameter}{distance} = distance \cdot \frac{angular\ diameter}{206,265"}$$

$$linear\ diameter = \left(7.8\times10^7\ km\right) \cdot \left(\frac{1"}{206,265"}\right) = 380\ km.$$

A feature on Mars that appears to be 1 second of arc when Mars is at 0.52 AU from Earth would have a size of 380 km.

2. The moon is at a distance of $3.84 \times 10^5$ km and completes an orbit in 27.32 days. Based on this information, compute the mass of Earth.

   Solution:
   We know
   the orbital distance of the moon = $a$ = $3.84 \times 10^5$ km and
   the orbital period of the moon = $P$ = 27.32 days.

   We can use the mass of a binary formula in *By the Numbers 8-4* on page 143 to find the mass of the Earth if we assume that the moon's mass is small compared to Earth's mass.

   $$M_{\text{solar masses}} = \frac{a_{AU}^3}{P_{\text{years}}^2}$$

   We need the orbital distance in units of astronomical units (1 AU = $1.5 \times 10^8$ km) and the orbital period in years (1 year = 365.25 days).

   $$a_{AU} = \frac{3.84 \times 10^5 \text{ km}}{1.5 \times 10^8 \frac{\text{km}}{\text{AU}}} = 2.56 \times 10^{-3} \text{ AU, and}$$

   $$P_{\text{years}} = \frac{27.32 \text{ days}}{365.25 \frac{\text{days}}{\text{years}}} = 7.48 \times 10^{-2} \text{ years}$$

   Now we can put these values into the mass formula

   $$M_{\text{solar masses}} = \frac{a_{AU}^3}{P_{\text{years}}^2} = \frac{\left(2.56 \times 10^{-3}\right)^3}{\left(7.48 \times 10^{-2}\right)^2} = \frac{1.68 \times 10^{-8}}{5.59 \times 10^{-3}} = 3.0 \times 10^{-6} \text{ solar masses}$$

   From Table A-6 on page 456 we find that the mass of the sun is approximately $2.0 \times 10^{30}$ kg.

   The mass of Earth then is equal to ($3.0 \times 10^{-6}$ solar masses) · ($2.0 \times 10^{30}$ kg) = $6.0 \times 10^{24}$ kg.

3. The graph on the right plots the escape velocity of each planet along the vertical axis and its surface temperature along the horizontal. The lines plotted in the figure are the average speeds of gas particles as a function of temperature for various gases. Which of the gases plotted in the diagram could be retained in the atmosphere of Mars?

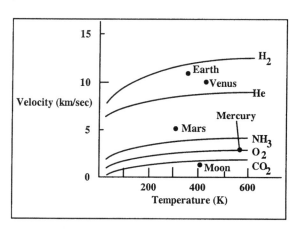

Solution:
From the graph we find that the escape velocity of Mars is about 5 km/s and that its surface temperature is about 300 K. The graph shows us that at 300 K the velocity of $H_2$ molecules is over 11 km/s. Since the average velocity of an $H_2$ molecule is greater than the escape velocity, $H_2$ molecules would quickly escape from Mars and would not be expected to be found in the Martian atmosphere. At 300 K the velocities of each material are; 11 k/m for $H_2$, 8 km/s for He, 4 km/s for $NH_3$, 2.5 km/s for $O_2$, and 2 km/s for $CO_2$. Since the escape velocity of Mars is about 5 km/s, any particle with a speed less than 5 km/s will not escape from Mars. <u>Therefore, the molecules that Mars can retain in its atmosphere are $CO_2$, $NH_3$, and $O_2$.</u>

## MORE MATH RELATED PROBLEMS
(Answers at the end of the chapter)

1. Deimos orbits Mars at a distance of $2.35 \times 10^4$ km in 1.26 days. Based on this information, what is the mass of Mars?

2. The smallest detail observable with Earth-based telescopes is about 1 second of arc in diameter? What size does this correspond to on the surface of Mercury at its closest approach to Earth? (Hint. At its closest approach to Earth, Mercury is at a distance of 0.61 AU.)

3. If the surface of Mercury has a temperature of 800 K, what wavelength does it radiate the most energy? Hint use *By the Numbers 6-1* on page 98.

## PRACTICE TEST
## Multiple Choice Questions

1. _____ formed as Mercury cooled and shrank.
   a. Shield volcanoes
   b. Coronae
   c. Rift valleys
   d. Ishtar Terra
   e. Lobate scarps

2. The moon stopped evolving early in its history because
   a. it was struck by a large object that fractured it.
   b. it did not retain its atmosphere.
   c. it is too small to have kept its internal heat.
   d. it did not have a metal core.
   e. it was too close to Earth.

3. The presence of folded mountain chains, rift valleys, and composite volcanoes is evidence
   a. that Earth differentiated.
   b. of outgassing.
   c. of plate tectonics.
   d. that Earth has a molten iron core.
   e. that Earth is geologically inactive.

4. In the development of a planet, the stage of _____ occurred when the interior of the planet became warm and heavy elements sank to the core, while lighter elements rose to the surface.
   a. differentiation
   b. cratering
   c. glaciation
   d. accretion
   e. flooding

5. _____ waves are seismic waves that do not travel through Earth's liquid core.
   a. P or pressure
   b. S or stress
   c. F or force
   d. G or gravitational
   e. E or erratic

6. Earth and the moon could not have condensed from the same materials because
   a. they have different densities and compositions.
   b. the moon does not have an atmosphere like Earth's.
   c. the moon does not have a magnetic field.
   d. Earth has a greater mass than the moon.
   e. Earth's interior is much hotter than the moons.

7. The graph to the right indicates the temperature of Earth's atmosphere as a function of altitude. Over what range in altitude does the temperature change the least?
   a. 0 km to 10 km
   b. 10 km to 50 km
   c. 50 km to 100 km
   d. 100 km to 150 km

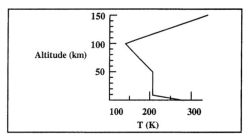

8. Besides Earth, which terrestrial planet or satellite of a terrestrial planet show evidence for the possible existence of liquid water flowing on its surface in the past?
   a. Venus
   b. the moon
   c. Mercury
   d. Mars
   e. Phobos

9. The surface temperature of Venus is very high because
   a. the interior of Venus is very hot.
   b. volcanoes on Venus erupt nearly continuously.
   c. Venus rotates in the opposite direction of the other planets.
   d. Venus has a very elliptical orbit.
   e. there is a large amount of carbon dioxide in Venus' atmosphere.

10. The graph on the right plots the escape velocity of each planet along the vertical axis and its surface temperature along the horizontal. The lines plotted in the figure are the average speeds of gas particles as a function of temperature for various gases. Which of the gases plotted in the diagram would not be expected to be found in the atmosphere of Venus?
    a. only $CO_2$
    b. only $NH_3$
    c. $CO_2$, $NH_3$, and $O_2$
    d. only $H_2$
    e. $H_2$ and He

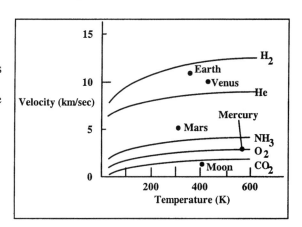

**Fill in the Blank Questions**

11. In the development of the moon, the stage of _____ occurred when molten rock flowed through fissures and filled deep basins.

12. The crust of Mars is believed to be much _____ than Earth's crust.

13. The _____ occurs because carbon dioxide is transparent to visible light and opaque to infrared radiation.

14. _____ occurs as rocks are heated and release gas rich in carbon dioxide, nitrogen, and water vapor.

15. Flow channels on _____ suggest it was once rich in water.

**True-False Questions**

16. The ozone layer on Earth is opaque to ultraviolet radiation.

17. The extreme size of volcanoes on Mars indicates that Mars has a much larger iron core than Earth.

18. The large mean density of Mercury supports the belief that Mercury has a large iron core.

19. Current evidence better supports the capture hypothesis than it does other hypotheses for the formation of the Earth-moon system.

20. Coronae on Venus are believed to be caused by rising convection currents in the interior of Venus.

## ADDITIONAL READING

Bullock, Mark A. and David H. Grinspoon "Global Climate Change on Venus." *Scientific American 280* (March 1999), p. 50.

Gurnis, Michael "Sculpting the Earth from the Inside Out." *Scientific American 284* (March 2001), p. 40.

Hoffman, Paul F. and Daniel P. Schrag "Snowball Earth." *Scientific American 284* (Jan. 2001), p. 68.

Talcott, Richard "Mars Returns to Glory." Astronomy 29 (July 2001), p. 48.

Zorpette, Glenn "Why Go to Mars?" *Scientific American 283* (March 2000), p. 40.

## ANSWERS TO MORE MATH RELATED PROBLEMS
1. $3.25 \times 10^{-7}$ solar masses $= 6.5 \times 10^{23}$ kg
2. 440 km
3. 3750 nm

## ANSWERS TO PRACTICE TEST

| | | | | | |
|---|---|---|---|---|---|
| 1. | e | 8. | d | 15. | Mars |
| 2. | c | 9. | e | 16. | T |
| 3. | c | 10. | d | 17. | F |
| 4. | a | 11. | flooding | 18. | T |
| 5. | b | 12. | stronger or thicker | 19. | F |
| 6. | a | 13. | greenhouse effect | 20. | T |
| 7. | b | 14. | Outgassing | | |

# CHAPTER 18

# WORLDS OF THE OUTER SOLAR SYSTEM

## UNDERSTANDING THE CONCEPTS

### 18-1  Jupiter

**Window on Science 18-1.  Science, Technology, and Engineering**

What is the difference between science and technology?

What is the value of science to humankind?

What is the central concept of this sub-section?

**The Interior**

What does the density of Jupiter tell us?

What do observations of Jupiter's heat flow reveal?

What is liquid metallic hydrogen?

How is Jupiter's magnetic field produced?

What is the central concept of this sub-section?

**Jupiter's Atmosphere**

What is the composition of Jupiter's atmosphere?

What is belt-zone circulation?

What was learned from the comet impacts on Jupiter in 1994?

What is the central concept of this sub-section?

**Jupiter's Ring**

What evidence supports the belief that the particles in Jupiter's ring are very small?

What is the Roche limit?

What evidence do we have that the rings are resupplied with dust?

What is the central concept of this sub-section?

**Jupiter's Family of Moons**

Which of the Galilean moons of Jupiter are suspected of containing oceans of water under a thick layer of ice (there are two of them)?

What causes the internal heat on Io that produces the volcanoes?

What is the central concept of this sub-section?

**A History of Jupiter**

How did Jupiter's composition come to resemble the composition of the sun, while Earth's is so different?

What is the central concept of this sub-section?

What are the fundamental ideas presented in this section?

### 18-2  Saturn

**Saturn the Planet**

Why are the belt-zone patterns less visible on Saturn than on Jupiter?

What does Saturn's density tell us?

What does the oblateness of Saturn tell us?

What is the central concept of this sub-section?

**Saturn's Rings and The Ice Rings of Saturn on pages 394 and 395**

What are the rings of Saturn made of?

Why do we know that the material in the rings must be replenished?

What causes the gaps in Saturn's rings?

What is the central concept of this sub-section?

**The Moons of Saturn**
>What is the atmosphere of Titan like?
>Why is Titan so cold?
>What is unusual about Enceladus?
>What is the central concept of this sub-section?

**Window on Science 18-2. Who Pays for Science**
>Why don't industries fund astronomical research?
>Who funds astronomical research?
>What is the central concept of this sub-section?

**The History of Saturn**
>How did gravitational collapse affect the formation of Saturn?
>Where does the material in Saturn's rings come from?
>What is the central concept of this sub-section?

What are the fundamental ideas presented in this section?

### 18-3 Uranus
**Uranus the Planet**
>Why doesn't Uranus contain liquid metallic hydrogen like Jupiter and Saturn?
>What is unusual about Uranus' rotation axis?
>Why does Uranus appear blue?
>What is the central concept of this sub-section?

**The Rings of Uranus and The Rings of Uranus and Neptune on page 400 and 401**
>What are the particles like that make up Uranus' rings?
>What does the narrowness of the rings tell us?
>What is the central concept of this sub-section?

**The Uranian Moons**
>Which of the Uranian moons shows evidence of flooding?
>What is the central concept of this sub-section?

**A History of Uranus**
>What may be the cause of Uranus' small size compared to Jupiter and Saturn?
>What might have caused the extreme tilt in Uranus' rotation axis?
>What is the central concept of this sub-section?

What are the fundamental ideas presented in this section?

### 18-4 Neptune
**Planet Neptune**
>Why is Neptune more blue than Uranus?
>What is the Great Dark Spot?
>What drives the atmospheric activity on Neptune?
>What is believed to produce the magnetic fields of Neptune and Uranus?
>What is the central concept of this sub-section?

**The Rings of Neptune and The Rings of Uranus and Neptune on page 400 and 401**
>How do we know that the rings of Neptune contain small dust grains?
>What is the central concept of this sub-section?

**The Moons of Neptune**
>What do the orbits of Triton and Nereid tell us?
>What evidence suggests that Triton has been active in the past?
>What is the central concept of this sub-section?

**A History of Neptune**
>Which planet is believed to have a history similar to Neptune?
>What is the central concept of this sub-section?

What are the fundamental ideas presented in this section?

**18-5  Pluto**

    **Pluto as a Planet**

        What is unique about Pluto's orbit?

        What is the composition of Pluto's atmosphere?

        What is the central concept of this sub-section?

    **The History of Pluto**

        What evidence do we have that Pluto is one of a group of similar objects in the outer solar system?

        What is the central concept of this sub-section?

What are the fundamental ideas presented in this section?


## KEY CONCEPTS

The Jovian planets have such thick atmospheres and vastly different compositions than the terrestrial planets that it is impossible to make a direct comparison between the two. Jupiter is the closest, largest and best studied of the Jovian planets and will serve as the standard for the comparative planetology of the Jovian planets. The major points on which this comparison is based are the structure of the interior and atmosphere, the strength of the magnetic field, and the structure of the rings and satellites of each planet.

All of the Jovian planets are of relatively low density, from about 0.7 $g/cm^3$ to about 1.7 $g/cm^3$. This implies that they contain a relatively small core of heavy elements, and large atmospheres of icy materials. Jupiter and Saturn are both believed to contain liquid metallic hydrogen. Liquid metallic hydrogen is an excellent conductor of electricity, and currents in the material are believed to produce the large magnetic fields associated with these planets. The interiors and atmospheres of Uranus and Neptune are very similar to each other. The data on both planets is limited but sufficient to begin to construct models of their interiors.

The rings of Saturn are spectacular and demonstrate that rings must be continually resupplied with material, presumable from material ejected from nearby moons. Saturn's rings are made of large icy blocks, ranging in size from dust to a few meters in diameter. The rings of the other Jovian planets are made of very small particles. We know the particles are small because the rings are effective at scattering light in a forward direction, something only very small particle do. Larger particles can exist in the rings, but there must be a relatively large number of the very small particles for the forward scattering of the light to be so strong.

The moons of the Jovian planets show us the importance of following the energy. Tidal heating appears to warm several of the Jovian moons. Io and Triton are hot enough in their interiors to produce volcanoes and/or geysers, both signs of active surfaces. Even Europa shows evidence that it has recently resurfaced, indicating that it has a relatively warm subsurface layer.

Our understanding of Pluto is very limited. There is very little information on Pluto or its moon Charon. We do know Pluto's mass, radius and density, which tells us that it is a mixture of rock and ice.

It is easy to get caught up in all of the neat facts that we have accumulated about the Jovian planets. The Voyager program, the Hubble Space Telescope, and the Galileo and Cassini missions have taught us a great deal about these bodies. Most important aspect of this chapter is how all of these observations and facts tell us something more about the manner in which nature operates. This chapter emphasizes how all of the neat little facts can be put together to tell us much more about the planets themselves and about how the physical universe works. What we learn in our backyard (i.e. our solar system) will help us better understand what happens throughout the universe.

## WEBSITES OF INTEREST

http://seds.lpl.arizona.edu/nineplanets/nineplanets/nineplanets.html   Great Solar System Site
http://www.jpl.nasa.gov/solar_system/                                 NASA Jet Propulsion Lab
http://www.jpl.nasa.gov/galileo/Jovian.html                           Galileo Mission Site
http://voyager.jpl.nasa.gov/                                          Voyager Mission Site
http://saturn.jpl.nasa.gov/index.cfm                                  Cassini-Huygens Mission Site

## QUESTIONS ON CONCEPTS

1. Why can Jovian planets have rings, while terrestrial planets cannot?

Solution:
The rings of the Jovian planets require the presence of the moons to provide material to resupply the rings. The rings should dissipate rather quickly and material is continually added from the nearby moons. The moons also help to define the orbits of the ring material. The terrestrial planets lack the moons required to maintain the ring material. In the end, it is the mass of the planet that allows them to have moons, and rings. Low mass planets cannot hold onto large number of moons or ring material. The massive Jovian planets have the gravitational energy to maintain a large number of moons that can produce a sufficient amount of material to maintain the ring structures.

2. Why do we believe that the interior of Jupiter contains liquid metallic hydrogen?

Solution:
Two observable properties of Jupiter provide clues about its interior. First, the average density of Jupiter tells us that Jupiter must be composed of light atoms and molecules. The average density clearly indicates that Jupiter cannot possess a massive iron core as found in the terrestrial planets. The other observable property is the strength of Jupiter's magnetic field. Jupiter's field is 10 times stronger than Earth's. The mechanism for producing magnetic fields requires a fluid material that is a great conductor of electricity. In Earth, it is the liquid iron core that produces the magnetic field. Jupiter's strong magnetic field and lack of an iron core means that another material must be highly conductive and capable of producing magnetic fields. Liquid metallic hydrogen is capable of doing just that. In the end though, it is computer models that we use to investigate the interior of Jupiter. Computer models based on Jupiter's density, mass, radius, and magnetic field indicate that it contains a significant amount of liquid metallic hydrogen. Note, we haven't sampled this material, we deduce its existence in Jupiter based on the density, magnetic field and computer models.

## WORKED EXAMPLES OF PROBLEMS REQUIRING MATHEMATICS

1. Pluto at its closest approach to Earth is 28.6 AU. At that point Pluto has an angular diameter of about 0.1 seconds of arc. Based on this information, what is the diameter of Pluto? (Hint 1 AU = $1.5\times10^8$ km).

Solution:
We know
the distance to Pluto = 28.6 AU = (28.6 AU) · ($1.5\times10^8$ km/AU) = $4.29\times10^9$ km,
the angular diameter of Pluto = 0.1 second of arc.

We can use the small angle formula in *By the Numbers 3-1* on page 32 to solve for the linear diameter of Pluto.

$$\frac{linear\ diameter}{distance} = \frac{angular\ diameter}{206,265"}$$

If we multiply both sides by the distance, we can solve for the linear diameter.

$$linear\ diameter = distance \cdot \frac{linear\ diameter}{distance} = distance \cdot \frac{angular\ diameter}{206,265"}$$

$$linear\ diameter = \left(4.29 \times 10^9\ km\right) \cdot \left(\frac{0.1}{206,265"}\right) = 2080\ km.$$

<u>Based on the given data the diameter of Pluto would be 2,080 km.</u>

2. The diagram below is meant to show the outline of a rotating planet.
   a. Draw a line that represents the rotation axis of the planet.
   b. Determine the oblateness of the ellipse drawn below.

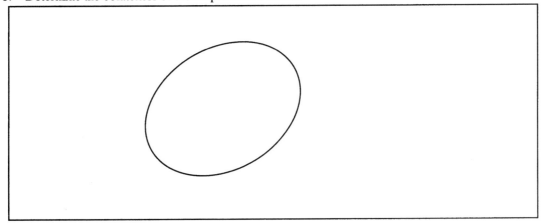

<u>Solution to part a:</u>
A rotating object will have its greatest diameter along its equator and its smallest diameter along its rotational axis. So the rotation axis is always the shortest diameter of the rotating object.

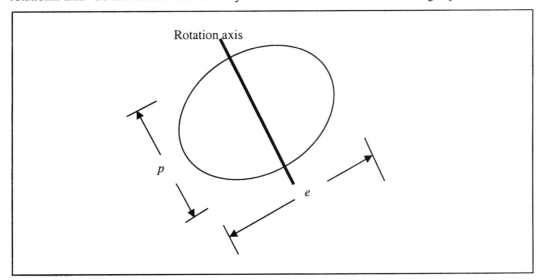

<u>Solution to part b:</u>
The oblateness of a planet is equal to the fraction by which the equatorial diameter exceeds the polar diameter. The equation for the oblateness is not given in the textbook, but can be deduced from its definition. In the figure above the polar diameter is the length labeled "*p*" and the equatorial diameter is the length labeled "*e*". Measuring these we find

$p \approx 3.2$ cm and
$e \approx 4.5$ cm.

Now we can calculate the oblateness.

$$Oblatness = \frac{e\text{-}p}{e} = \frac{(4.5-3.2)\,\text{cm}}{4.5\,\text{cm}} = \frac{1.3}{4.5} = 0.29$$

<u>The oblateness of this planet is 0.29.</u>

3. The orbital period of Pluto is 288 years and its average distance from the sun is $5.9 \times 10^9$ km.
   a. How many kilometers along its orbit does Pluto move in 24 hours?
   b. How many seconds of arc would Pluto appear to move to an observer on Earth (take the distance to be $5.9 \times 10^9$ km).

<u>Solution to part a:</u>
We know
the orbital distance of Pluto = $r = 5.9 \times 10^9$ km,
the orbital period of Pluto = $P$ = 288 years = (288 years) · (365.25 days/year) = $1.05 \times 10^5$ days.

The distance that Pluto travels in one day is equal to the circumference of Pluto's orbit divided by the time it takes Pluto to orbit the sun in units of days.

$$\text{distance traveled per day} = \frac{\text{circumference of orbit}}{\text{orbital period in days}} = \frac{2\pi r}{P_{\text{days}}} = \frac{2\pi \cdot \left(5.9 \times 10^9\ \text{km}\right)}{1.05 \times 10^5\ \text{days}} = 3.52 \times 10^5\ \text{km/day}$$

<u>Pluto moves approximately $3.52 \times 10^5$ km in 24 hours.</u>

<u>Solution to part b:</u>
We know
in one day Pluto moves a distance = $3.52 \times 10^5$ km and
the distance from Earth to Pluto at this time is = $5.9 \times 10^9$ km.

We can use the small angle formula in *By the Numbers 3-1* on page 32 to find the angular distance traveled by Pluto in one day.

$$\frac{angular\ distance}{206,265"} = \frac{linear\ distance}{distance}$$

If we multiply both sides by 206,265" we can solve for the angular distance Pluto appears to travel.

$$angular\ distance = 206,265" \cdot \frac{linear\ distance}{distance} = \left(206,265"\right) \cdot \left(\frac{3.52 \times 10^5\ \text{km}}{5.9 \times 10^9\ \text{km}}\right) = 12.3"$$

<u>Pluto would appear to move approximately 12.3 seconds of arc per day.</u> This solution ignores the motion of Earth and the parallax that produces.

4. Iapetus orbits a planet at a distance of $3.56 \times 10^6$ km and completes one orbit in 79.3 days. What is the mass of the planet that Iapetus orbits?

Solution:
We know
1 year = 365.25 days,
1 AU = $1.5 \times 10^8$ km,
the orbital period = $P$ = 79.3 days = (79.3 days)÷(365.25 days/year) = 0.217 years. and
the orbital distance = $a$ = $3.56 \times 10^6$ km = $(3.56 \times 10^6$ km)÷($1.5 \times 10^8$ km/AU) = 0.0237 AU.

We can use Newton's and Kepler's laws from *By the Numbers 8-4* on page 143 to find the mass of the planet.

$$M = \frac{a_{AU}^3}{P_{years}^2} = \frac{(0.0237 \text{ AU})^3}{(0.217 \text{ years})^2} = \frac{1.34 \times 10^{-5}}{4.71 \times 10^{-2}} = 2.84 \times 10^{-4} \text{ solar masses}$$

Now 1 solar mass = $2.0 \times 10^{30}$ kg, so

$M$ = ($2.84 \times 10^{-4}$ solar mass) · ($2.0 \times 10^{30}$ kg/solar mass) = $5.68 \times 10^{26}$ kg.

The mass of the planet is $5.68 \times 10^{26}$ kg. Iapetus orbits the planet Saturn.


## MORE MATH RELATED PROBLEMS
(Answers at the end of the chapter)

1. Larissa orbits Neptune at a distance of $7.36 \times 10^4$ km with a period of 0.554 days. What is the mass of Neptune based on these observations?

2. The table below contains orbital data for several planetary moons. Complete the table.

| Object | Orbital Distance (km) | Orbital Period (days) | Orbital Distance (AU) | Orbital Period (years) | Planet's Mass (solar mass) | Planet's Mass (kg) |
|---|---|---|---|---|---|---|
| A | $1.28 \times 10^5$ | 0.294 | | | | |
| B | $2.38 \times 10^5$ | 1.37 | | | | |
| C | $1.21 \text{x} \times 10^7$ | 12.84 | | | | |
| D | $3.55 \times 10^5$ | 5.88 | | | | |
| E | $2.35 \times 10^4$ | 1.262 | | | | |
| F | $2.24 \times 10^7$ | 692 | | | | |
| G | $1.48 \times 10^6$ | 21.28 | | | | |
| H | $5.25 \times 10^4$ | 0.333 | | | | |

3. Stephano is a moon of Uranus at an orbital distance of $7.9 \times 10^6$ km. Given that the diameter of Uranus is $5.1 \times 10^4$ km, what would the angular diameter of Uranus be when viewed from Stephano?

4. Cordelia is a moon of Uranus at an orbital distance of 49,800 km. Given that the diameter of Uranus is $5.1 \times 10^4$ km, what would the angular diameter of Uranus be when viewed from Cordelia?

5. Jupiter orbits the sun in 11.87 years and is located 5.20 AU from the sun.
   a. In kilometers, how far does Jupiter move along its orbit in 24 hours?
   b. If Jupiter is at a distance of 4.5 AU from Earth, what angular distance on the sky will it appear to move in 24 hours?

**PRACTICE TEST**
**Multiple Choice Questions**

1. There would be no rings around Saturn if
   a. Saturn had a larger density.
   b. Saturn had a larger radius.
   c. Saturn had a smaller radius.
   d. Saturn was closer to the sun.
   e. Saturn didn't have moons.

2. How can Titan keep an atmosphere when it is only slightly larger than airless Mercury?
   a. Titan has a much stronger magnetic field than Mercury.
   b. The atmosphere of Titan is resupplied by Saturn.
   c. Titan is much colder than Mercury.
   d. Titan's rotation is much slower than Mercury's.
   e. Titan is currently outgassing and forming its atmosphere.

3. The region of Jupiter that is responsible for producing the magnetic field is composed of
   a. liquid metallic hydrogen.
   b. molten iron.
   c. methane.
   d. solid iron.
   e. various organic compounds.

4. The particles in Jupiter's ring are
   a. about the same size as the wavelengths of visible light.
   b. about the size of tennis balls.
   c. about the size of basketballs.
   d. pieces of captured comets.
   e. material from eruptions of volcanoes on Io.

5. Besides Earth, volcanic eruptions have been observed on
   a. Io and Callisto.
   b. Titan and Iapetus.
   c. Deimos and Charon.
   d. Triton and Io.
   e. Ganymede and Charon.

6. Why are the belts and zones on Saturn less distinct than those on Jupiter?
   a. The belts and zones on Saturn are suppressed by Saturn's stronger magnetic field.
   b. The belts and zones on Saturn occur much deeper, below a layer of methane haze.
   c. The belts and zones on Saturn are suppressed because Saturn rotates faster.
   d. The atmosphere of Saturn contains a larger amount of hydrogen which inhibits belts and zones.
   e. Belts and zones do not exist in Saturn's atmosphere.

7.  Uranus is unique in that
    a.  it orbits in the opposite direction of the other planets.
    b.  its orbit is highly inclined relative to the ecliptic.
    c.  its rotation axis is nearly in the plane of its orbit.
    d.  it is the only Jovian planet without rings.
    e.  it is the only Jovian planet without a moon.

8.  The graph on the right plots the escape velocity of several solar system objects along the vertical axis and the surface temperature along the horizontal. The lines plotted in the figure are the average speeds of gas particles as a function of temperature for various gases. Which of the gases plotted in this diagram could be retained by Ganymede?
    a.  $O_2$ and $CO_2$
    b.  only $NH_3$
    c.  $CO_2$, $NH_3$, and $O_2$
    d.  all of them
    e.  None of them

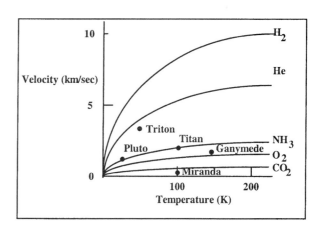

9.  The heating of the interior of Io is believed to be caused by
    a.  radio active decay.
    b.  infrared radiation from Jupiter.
    c.  the solar wind.
    d.  their magnetic fields.
    e.  tidal friction.

10. The rings of all the Jovian planets.
    a.  are composed of icy particles.
    b.  are composed of rocky particles about the size of a tennis ball.
    c.  are inside the planets Roche lobe.
    d.  formed at the time the planet formed.
    e.  are composed of material captured from passing comets.

**Fill in the Blank Questions**

11. _____ in the atmosphere of Neptune absorbs red light giving Neptune a blue color.

12. The rings of _____ were discovered during an occultation of a star.

13. The atmosphere of _____ contains mostly nitrogen with traces of methane.

14. _____ satellites keep the F ring narrow.

15. _____ is a moon of Saturn that may have organic particles on its surface as a result of the interaction of sunlight with methane in its upper atmosphere.

**True-False Questions**

16. Belt and zone circulation on Jupiter is caused by Jupiter's magnetic field.

17. The old surfaces of icy satellites appear dark in color with many impact craters.

18. Europa has few craters because it keeps one face always pointed toward Jupiter, which screens it from incoming meteoroids.

19. Pluto's density is 1.8 g/cm$^3$. This implies that Pluto is about 50% water and 50% rocky material.

20. The excess heat produced by Jupiter and Saturn is the result of radioactive decay.


## ADDITIONAL READING

Beatty, J. Kelly "Pluto Reconsidered." *Sky and Telescope 97* (May 1999), p. 48.

Carroll, Michael "Europa: Distant Ocean, Hidden Life?" *Sky and Telescope 94* (Dec. 1997), p. 50.

Elloit, James L. "The Warming Wisps of Triton." *Sky and Telescope 97* (Feb. 1999), p. 42.

Johnson, Torrence V. "The Galileo Mission to Jupiter and Its Moons." *Scientific American 283* (Feb. 2000), p. 40.

Sincell, Mark "Switched at Birth." *Astronomy 28* (March 2000), p. 48.

Talcott, Richard "A Giant Awakens Cassini." *Astronomy 29* (April 2001), p. 28.


## ANSWERS TO MORE MATH RELATED PROBLEMS

1. $5.09 \times 10^{-5}$ solar masses $= 1.02 \times 10^{26}$ kg

2.

| Object | Orbital Distance (km) | Orbital Period (days) | Orbital Distance (AU) | Orbital Period (years) | Planet's Mass (solar mass) | Planet's Mass (kg) |
|---|---|---|---|---|---|---|
| A | $1.28 \times 10^5$ | 0.294 | $8.53 \times 10^{-4}$ | $8.05 \times 10^{-4}$ | $9.59 \times 10^{-4}$ | $1.92 \times 10^{27}$ |
| B | $2.38 \times 10^5$ | 1.37 | $1.59 \times 10^{-3}$ | $3.75 \times 10^{-3}$ | $2.84 \times 10^{-4}$ | $5.68 \times 10^{26}$ |
| C | $1.21 \times 10^7$ | 1,284 | $8.07 \times 10^{-2}$ | 3.50 | $4.27 \times 10^{-5}$ | $8.55 \times 10^{25}$ |
| D | $3.55 \times 10^5$ | 5.88 | $2.37 \times 10^{-3}$ | $1.61 \times 10^{-2}$ | $5.11 \times 10^{-5}$ | $1.02 \times 10^{26}$ |
| E | $2.35 \times 10^4$ | 1.262 | $1.57 \times 10^{-4}$ | $3.46 \times 10^{-3}$ | $3.22 \times 10^{-7}$ | $6.44 \times 10^{23}$ |
| F | $2.24 \times 10^7$ | 692 | 0.149 | 1.89 | $9.28 \times 10^{-4}$ | $1.86 \times 10^{27}$ |
| G | $1.48 \times 10^6$ | 21.28 | $9.87 \times 10^{-3}$ | $5.83 \times 10^{-2}$ | $2.83 \times 10^{-4}$ | $5.66 \times 10^{26}$ |
| H | $5.25 \times 10^4$ | 0.333 | $3.50 \times 10^{-4}$ | $9.12 \times 10^{-4}$ | $5.16 \times 10^{-5}$ | $1.03 \times 10^{26}$ |

3. 1,300 seconds of arc = 22 minutes of arc, or about two-thirds the angular diameter of the moon as viewed from Earth.

4. 211,000 seconds of arc = 3,520 minutes of arc = 58.7 degrees.

5. a.  $1.13 \times 10^6$ km/day
   b.  345 seconds of arc = 5.74 minutes of arc

## ANSWERS TO PRACTICE TEST

| | | |
|---|---|---|
| 1. e | 8. a | 15. Titan |
| 2. c | 9. e | 16. F |
| 3. a | 10. c | 17. T |
| 4. a | 11. Methane | 18. F |
| 5. d | 12. Uranus | 19. T |
| 6. b | 13. Pluto | 20. F |
| 7. c | 14. Shepard | |

CHAPTER 19

# METEORITES, ASTEROIDS, AND COMETS

## UNDERSTANDING THE CONCEPTS

### 19-1  Meteorites
**Inside Meteorites**
What are the characteristics of the three types of meteorites?
What are Widmanstätten patterns and where are they found?
What are chondrites?
What is a chondrule?
What are carbonaceous chondrites?
What is the central concept of this sub-section?
**The Origin of Meteors and Meteorites**
What is a meteor shower?
Why do meteor showers occur?
What objects produce the majority of the meteors we see?
What objects produce the majority of the meteorites we find?
What is the central concept of this sub-section?
**Window on Science 19-1.  Enjoying the Natural World**
How can science increase our enjoyment of nature?
What is the central concept of this sub-section?

What are the fundamental ideas presented in this section?

### 19-2  Asteroids
**Properties of Asteroids**
Why do astronomers believe that most asteroids are not spherical?
What does the density of Mathilde tell us about asteroids?
What evidence tells us that collisions have occurred frequently among the asteroids?
What are the characteristics of the three types of asteroids?
How are the three types of asteroids and three types of meteorites related?
What is the central concept of this sub-section?
**The Origin of Asteroids**
What is our current hypothesis concerning the formation of the asteroids?
What is the central concept of this sub-section?

What are the fundamental ideas presented in this section?

### 19-3  Comets
**Properties of Comets and Comet Observations on pages 422 and 423**
What are the properties of the two kinds of tails that comets posses?
What do the tails of comets tell us?
What is the nucleus of a comet made of and what is its size?
What effect does the solar wind play on comets?
What is the central concept of this sub-section?

### The Geology of Comet Nuclei
What is the composition of cometary nuclei?

What have we learned about the structure of cometary nuclei from the density of comets?

Why are dirty snowballs or icy mud-balls misleading analogies of a comet?

What is the central concept of this sub-section?

### The Origin of Comets
What are the orbital characteristics of the comets with periods greater than 200 years?

What are the orbital characteristics of the comets with periods less than 200 years?

What is the Oort cloud?

What is the Kuiper belt?

How is Pluto related to the Kuiper belt?

How did the objects in the Oort cloud get there?

What is the central concept of this sub-section?

### Impacts on Earth
What evidence indicates that an impact by a comet or asteroid is responsible for the extinction of the dinosaurs?

What sequence of events might occur if Earth were struck by a large asteroid or comet?

What is the central concept of this sub-section?

What are the fundamental ideas presented in this section?


## KEY CONCEPTS

The goal of this chapter is to develop an understanding of how meteorites, asteroids, and comets can be used to better understand the physical processes of our solar system and its formation.

Meteorites are objects that we can hold - this is very unique in astronomy. Meteorites provide an opportunity to accurately determine the chemical composition and physical properties of material that has spent most of its life outside Earth's atmosphere. The meteorites are used as the base for a comparative study of the asteroids and comets in the remainder of the chapter.

The evidence strongly suggests that most meteorites are composed of asteroidal material, and the study of the composition of meteorites tells us about the formation of the asteroids. This study leads to the conclusion that many of the planetesimals that formed in the asteroid belt were big enough to be geologically active for a significant time and that some of these planetesimals differentiated.

The study of the spectra and images of comets and cometary nuclei indicates that they are modeled fairly well by the dirty snowball model modified to include large cavities, pits, and trenches. Additionally, most of the comets may have originally formed in the Kuiper belt, a region just beyond the orbit of Uranus. There is a strong enough relationship in terms of composition between comets, Pluto and several of the satellites of Uranus and Neptune to suggest that these objects may have formed in the same region. This leads to the suggestion that Pluto may be one of the largest planetesimals left that formed the comets of the Kuiper Belt.

Most of the long period comets appear to originate in the Oort cloud. Most of these comets are believed to have formed in the vicinity of the Jovian planets and then launched to the Oort clouds by gravitational interactions with the young giant planets.

## QUESTIONS ON CONCEPTS

1. The diagram at the right is a plot of an asteroid's reflected brightness versus it ultraviolet minus visual color (U–V color). In this diagram, which of the asteroids is most likely an S-type asteroid?

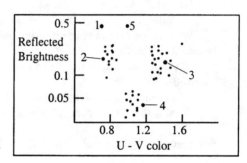

Solution:
The three types of asteroids are defined based on their U–V color and their reflected brightness, also knows as their albedo. When these two observational properties are plotted on a graph, the asteroids fall into three well defined groups, see Figure 19-6. The M-type asteroids have small U–V colors and fairly large albedos. The S-type asteroids have large U–V colors and fairly large albedos. The M-type asteroids have low to moderate U–V colors and low albedos. We are asked to determine which object is most likely an S-type asteroid. So we want the object with a relatively large U–V color and a relatively large albedo (i.e. reflected brightness). The labeled object with the characteristics like the S-type asteroids is Object 3.

2. What evidence do we have that an impact from an asteroid may have caused the mass extinction of the dinosaurs?

Solution:
There are four observations that lead to the conclusion that the impact of a meteor may have caused the extinction of the dinosaurs. First, we find an overabundance of iridium in the layers of soil formed at that time. Most of the other layers of Earth's soils contain very little iridium. Secondly, abundance analysis of meteorites tells us that they meteorites have a greater abundance of iridium than Earth's crust. The impact of a large meteorite would distribute material across Earth's surface and add its material to the soil layers. This could account for an increased abundance of iridium in layers of soil put down in the same time period of a major asteroid impact. Third, we find large amounts of soot from fires in the soil layers formed at that time. A major impact is expected to loft large amounts of molten rock into the atmosphere that would fall across the surface of Earth and initiate forest fires in many locations. These fires would darken the sky and produce a layer of soot that would cover most of Earth's solid surface. Finally, the Chicxulub meteor crater dates to this same time period. The crater shows the shocked quartz indicative of an impact and dates to about 65 million years. Taken together, these observations suggest that an asteroid impact on Earth may have been responsible for the extinction of the dinosaurs.

**WORKED EXAMPLES OF PROBLEMS REQUIRING MATHEMATICS**

1.  What is the orbital velocity of a meteoroid that orbits the sun at a distance of 1.3 AU?

    Solution:
    This problem can be completed in two different ways. One is to use the circular velocity formula from *By the Numbers 4-1* on page 62. This will require converting the distance into meters and looking up the value of the sun's mass and the gravitational constant, $G$. The second method is to find the period of the object's orbit from Kepler's law, $P^2 = a^3$, and then divide the circumference of the orbit by the orbital period. Both solutions are presented, and either solution will work.

    Solution using Circular Velocity:
    We know
    the orbital distance $= a = 1.3$ AU $= (1.3$ AU$) \cdot (1.5 \times 10^{11}$ m$) = 1.95 \times 10^{11}$ m,
    the mass of the sun $= M = 2.0 \times 10^{30}$ kg, and
    the gravitational constant $= G = 6.67 \times 10^{-11}$ m$^3$/(kg·s$^2$).

    The formula for the circular velocity is given in *By the Numbers 4-1* on page 62.

    $$V_c = \sqrt{\frac{GM}{a}} = \sqrt{\frac{\left(6.67 \times 10^{-11} \frac{m^3}{kg \cdot s^2}\right) \cdot \left(2.0 \times 10^{30} \text{ kg}\right)}{1.95 \times 10^{11} \text{ m}}} = \sqrt{\frac{1.33 \times 10^{20}}{1.95 \times 10^{11}}} = 2.6 \times 10^4 \text{ m/s} = 26 \text{ km/s}$$

    The orbital velocity of the object is 26 km/s.

    Solution using Kepler's law:
    We know
    the orbital distance $= a = 1.3$ AU $= (1.3) \cdot (1.5 \times 10^8$ km$) = 1.95 \times 10^8$ km.

    Kepler's third law can be used to find the period of the orbit.

    $$P^2_{years} = a^3_{AU} = (1.3 \text{ AU})^3 = 2.20$$

    $$P_{years} = \sqrt{P^2_{years}} = \sqrt{2.20} = 1.48 \text{ years}$$

    $$P_{seconds} = (1.48 \text{ years}) \cdot \left(365.25 \tfrac{days}{year}\right) \cdot \left(24 \tfrac{hours}{day}\right) \cdot \left(3600 \tfrac{seconds}{hour}\right) = 4.68 \times 10^7 \text{ s}$$

    The orbital speed $= V_c =$ the circumference of the orbit divided by the orbital period.

    $$V_c = \frac{Circumference}{Period} = \frac{2\pi a}{P} = \frac{2\pi \cdot \left(1.95 \times 10^8 \text{ km}\right)}{4.68 \times 10^7 \text{ s}} = 26 \tfrac{km}{s}$$

    The orbital velocity of the object is 26 km/s, exactly as we found using the first method.

2.  If you observe a comet at a distance of 1.6 AU from Earth and it has a visible tail 2° long, what is the length of the tail in km?

    Solution:
    We know
    the distance to the comet = 1.6 AU = (1.6 AU) · (1.5×10$^8$ km/AU) = 2.4×10$^8$ km,
    the angular length of the tail = 2° = (2°)· (3600 seconds of arc per °) = 7,200 seconds of arc.

    We can use the small angle formula in *By the Numbers 3-1* to find the length of the tail.

    $$\frac{linear\ length}{distance} = \frac{angular\ length}{206,265"}$$

    If we multiply both sides by the distance, we can solve for the linear length.

    $$linear\ length = (distance)\cdot\left(\frac{angular\ length}{206,265"}\right) = (2.4\times10^8\ km)\cdot\left(\frac{7,200"}{206,265"}\right) = 8.4\times10^6\ km$$

    The tail would be 8.4×10$^6$ km long assuming that it is perpendicular to our line of sight. If the tail is pointed away from us or toward us a little, it would be longer than 8.4×10$^6$ km.

3.  A near Earth asteroid is observed at a distance of 4.2×10$^6$ km and has a circular shape with an angular diameter of 0.1 seconds of arc. What is the linear diameter of this asteroid?

    Solution:
    We know
    the distance to the asteroid = 4.2×10$^6$ km and
    the angular diameter of the asteroid = 0.1 seconds of arc.

    We can use the small angle formula from *By the Numbers 3-1* on page 32 to find the linear diameter.

    $$\frac{linear\ diameter}{distance} = \frac{angular\ diameter}{206,265"}$$

    We multiply both sides by the distance and solve for the linear diameter.

    $$linear\ diameter = (distance)\cdot\left(\frac{angular\ diameter}{206,265"}\right) = (4.2\times10^6\ km)\cdot\left(\frac{0.1"}{206,265"}\right) = 2.0\ km$$

    The asteroid has a diameter of 2.0 km.

**MORE MATH RELATED PROBLEMS**
    (Answers at the end of the chapter)

1.  A comet is found with an orbital period of 76 years. What is the average orbital distance of this object?

2.  An asteroid is found with an average orbital distance of 3.2 AU, what is its orbital period?

3.  An asteroid with a mass of 2×10$^{18}$ kg is fragmented into pieces that average 800 kg. About how many pieces are there?

4. A near Earth asteroid is observed at a distance of $8.4 \times 10^5$ km and has a circular shape with an angular diameter of 0.1 seconds of arc. What is the linear diameter of this asteroid?

5. A near Earth asteroid is observed at a distance of $1.5 \times 10^6$ km and has an oval shape. The long axis of the asteroid is 0.4 seconds long and the short axis 0.2 seconds of arc long. What are the dimensions of the asteroid in kilometers?


**PRACTICE TEST**
**Multiple Choice Questions**

1. The iron meteorites
   I.    appear to be fragments from the iron core of a planetesimal.
   II.   were most likely once part of a comet.
   III.  show Widmanstätten patterns.
   IV.   are very heavy for their size and have dark irregular surfaces.

   a.   I & III
   b.   II & III
   c.   I & IV
   d.   II & IV
   e.   I, III, & IV

2. Meteorites seem to be primarily composed of material very similar to the material
   a.   in comets.
   b.   in moons of Neptune and Pluto.
   c.   in asteroids.
   d.   in the Oort cloud.
   e.   in the Kuiper belt.

3. The _____-type asteroids are generally found in the outer portion of the asteroid belt and are very dark.
   a.   C
   b.   B
   c.   M
   d.   S
   e.   X

4. The density of the comets measured to date tells us that a comet' nucleus is a
   a.   tightly packed ball of about equal parts ices and rocky material.
   b.   tightly packed ball of about 90% ices and 10% rocky material.
   c.   tightly packed ball of about 10% ices and 90% rocky material.
   d.   loose collection of about equal parts ices and rocky material, with trenches and cavities.
   e.   loose collection of muddy ice with a smooth crust of mostly frozen ices.

5. What evidence do we have that the asteroids fragmented?
   a.   The three types of asteroids appear similar to the crust, mantle, and core of a differentiated object.
   b.   Several of the larger asteroids show large impacts and non-spherical shapes.
   c.   Meteorites are found that are mostly stony, stony and iron mix, and mostly iron in composition.
   d.   Asteroids are found orbiting each other.
   e.   All of the above

6.  What evidence suggests that a large impact by an asteroid may have caused the mass extinctions at the end of the Cretaceous period 65 million years ago?
    I.   The over abundance of iridium found in a single layer throughout Earth's crust.
    II.  The discovery of the Chicxulub crater near the Yucatan peninsula.
    III. Soot from fires found in the layers of clay laid down at the end of the Cretaceous period.
    IV.  The discovery of fossils with small fragments of stony-iron meteorite embedded in the bones.

    a.  I, II, III, & IV
    b.  II, III, & IV
    c.  I, II, & IV
    d.  I, II, & III
    e.  I & II

7.  The M-type asteroids are believed to be similar in composition to
    a.  iron meteorites.
    b.  chondritic meteorites.
    c.  achondritic meteorites.
    d.  material in the meteor showers.
    e.  objects in the Kuiper belt.

8.  The diagram at the right is a plot of an asteroid's reflected brightness versus it ultraviolet minus visual color (U–V color). In this diagram, which of the asteroids is most likely an M-type asteroid?
    a.  Object 1
    b.  Object 2
    c.  Object 3
    d.  Object 4
    e.  Object 5

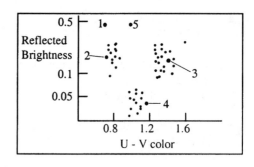

9.  The Kuiper belt
    I.   contains mostly C type asteroids.
    II.  is located between Mars and Jupiter.
    III. is where most of the meteorites originated.
    IV.  is inclined by nearly 90 degrees to the ecliptic.

    a.  I, II, & II
    b.  II, III & IV
    c.  II & III
    d.  I, II, III, & IV
    e.  none of them

10. An asteroid has an orbital period around the sun of 2.83 years. How far from the sun is this asteroid?
    a.  2.83 AU
    b.  2.0 AU
    c.  1.89 AU
    d.  8.0 AU
    e.  1.68 AU

**Fill in the Blank Questions**

11. _____ are round bits of glass found in some stony meteorites.

12. _____ are important in our understanding of solar system objects because we can take them into the laboratory and accurately determine their composition and analyze how Earth's atmosphere affected their surfaces.

13. _____ patterns are found only in iron meteorites.

14. If an asteroid _____, it could produce an iron core and a silicate mantle.

15. The long period comets, with periods in excess of 200 years, originate in the _____ _____.

**True-False Questions**

16. The gas tail of a comet always trails behind the comet along the orbital path.

17. A meteor shower is produced when Earth passes through the orbital path of a comet.

18. The density of Comet Halley was found to be about 0.2 $g/cm^3$. This implies that Comet Halley is tightly packed consisting of about 50% ices and 50% rocky material.

19. The M-type asteroids appear to be composed of carbonaceous chondrites.

20. Long period comets are believed to originate in the Oort cloud.


**ADDITIONAL READING**

Asphaug, Erik "The Small Planets." *Scientific American 283* (May 2000), p. 46.

Erwin, Douglas H. "The Mother of Mass Extinctions." *Scientific American 277* (July 1996), p. 72.

Gehrels, Tom "Collisions with Comets and Asteroids." *Scientific American 277* (March 1996), p. 54.

Hartmann, William K. "The Great Solar System Revision." *Astronomy 26* (Aug. 1998), p. 40.

Verschuur, Gerrit L. "Impact Hazards: Truth and Consequences." *Sky and Telescope 95* (June. 1998), p. 26.

Weissman, Paul " The Oort Cloud." *Scientific American 279* (Sept. 1998), p. 84.

**ANSWERS TO MORE MATH RELATED PROBLEMS**
1. 18 AU
2. 5.7 years
3. $2.5 \times 10^{15}$ pieces
4. 0.41 km = 410 m
5. 2.9 km×1.5 km


**ANSWERS TO PRACTICE TEST**

| | | |
|---|---|---|
| 1. e | 8. b | 15. Oort cloud |
| 2. c | 9. e | 16. F |
| 3. a | 10. b | 17. T |
| 4. d | 11. Chondrules | 18. F |
| 5. e | 12. Meteorites | 19. F |
| 6. d | 13. Widmanstätten | 20. T |
| 7. a | 14. differentiated | |

# C H A P T E R   2 0

# LIFE ON OTHER WORLDS

## UNDERSTANDING THE CONCEPTS

**20-1  The Nature of Life**
  **Window on Science 20-1.  The Nature of Scientific Explanation**
     Scientific explanations are based on two fundamental rules, what are they?
     Why are scientific explanations given so much weight?
     What is the central concept of this sub-section?
  **The Physical Basis of Life**
     Which atom is the physical basis for life?
     What property of carbon makes it perhaps the best basis for life?
     What is the central concept of this sub-section?
  **Information Storage and Duplication and DNA: The Code of Life on pages 434 and 435**
     Which molecules store the information that guides all the processes in an organism?
     What part of the cell contains the DNA?
     What is a gene?
     What is the central concept of this sub-section?
  **Modifying the Information**
     Describe what is meant by natural selection?
     What is the central concept of this sub-section?

  What are the fundamental ideas presented in this section?

**20-2  The Origin of Life**
  **The Origin of Life on Earth**
     How old are the oldest bacteria fossils?
     When did a wide diversity of complex forms appear?
     What is the significance of the Miller experiment?
     What is meant by the term primordial soup?
     What locations are expected to have been the birthplace of early complex molecules?
     What are stromatolites?
     What is the central concept of this sub-section?
  **Geologic Time**
     What is the name of the period that includes the first anthropoids and humans?
     What is the central concept of this sub-section?
  **Life in Our Solar System**
     Why is water important for the evolution of life?
     Which planets and/or moons within the solar system might have suitable conditions for life?
     What is the central concept of this sub-section?
  **Life in Other Planetary Systems**
     What is the life zone?
     What factors affect the size of the life zone?
     Stars of which spectral types are most likely to have planets on which life can evolve?
     What is the central concept of this sub-section?

  What are the fundamental ideas presented in this section?

### 20-3 Communication with Distant Civilizations
#### Travel Between the Stars
Why is travel between the stars nearly impossible?

What is the central concept of this sub-section?
#### Window on Science 20-2. UFOs and Space Aliens
Why can't we use reports of UFO sightings as evidence for the evolution of life in other star systems?

What is the central concept of this sub-section?
#### Radio Communication
Why is communication even at the speed of light problematic?

What two wavelengths might be best to search for signals from intelligent life in other star systems?

What is the water hole?

What is SETI?

What is the central concept of this sub-section?
#### How Many Inhabited Worlds?
What is the Drake equation?

What is the central concept of this sub-section?

What are the fundamental ideas presented in this section?

## KEY CONCEPTS

The key concept of this chapter is that life in our galaxy is highly probable; however, it will be very difficult to detect and to communicate with it. The first two sections of the chapter describe the bases of life and how it evolved on our planet. This provides a background for a look at where other life might be found and how we might detect it.

Life in our solar system beyond Earth has not been positively ruled out, but the odds are very small. Our understanding of the lives of stars allows us to determine the types of stars around which life has the best chances of evolving. This has helped us limit our search for star systems that might have suitable planets and might harbor life in some form.

The final section demonstrates that interstellar travel is really unfeasible and the only way to find and communicate with an alien world will be through radio waves. While the radio spectrum is large there are some frequencies that might provide a better place to search for the existence of life around other stars and to transmit our existence.

## WEBSITES OF INTEREST

| | |
|---|---|
| http://exoplanets.org/index.html | Extra-solar Planets site |
| http://www.seti-inst.edu/ | SETI Home Page |
| http://origins.jpl.nasa.gov/education/openhouse/openhouseidx.html | JPL Site, well organized |
| http://sci2.esa.int/specialevents/lifeinuniverse/ | European Space Agency site |

**QUESTIONS ON CONCEPTS**

1. What is natural selection?

   Solution:
   Natural selection is the process by which biological organisms evolve. Organisms that can live long enough to reproduce will extend the genetic information they carry to the next generation. Those organisms that do not live long enough to reproduce will kill the genetic information they carry. If the environmental changes, some organisms may not survive to reproduce, while others, that are only slightly different will. That slight difference is an advantage in the new environment and is passed on to subsequent generations. The environment changes might be very gradual, but the result is the same.

   A mutation may occur in the DNA, which can have one of three effects on the organism. The mutation may, 1) have little effect on the organism's ability to live long enough to reproduce, 2) inhibit the organism's ability to live to reproduce, or 3) make it more successful at reproducing. Benign changes will be passed on without effect, though a possible change to the environment could make this change important in determining whether the organism continues to pass on its genetic information. If the change is malevolent then the organism and its genetic information will die out, and this dangerous mutation will not be passed on. A benevolent mutation will aid the organism in its survival and increase the likelihood that the benefit to survival is passed on.

   Natural selection is a statement that simply says that those with the ability to reproduce will pass on their genetic material, and those that are not suited for their environment will die out.

2. What is the difference between chemical evolution and biological evolution?

   Solution:
   Chemical evolution is the random linking of elements and molecules to form larger stable molecules. The molecules will both combine and break apart randomly, and are incapable of reproducing themselves. The formation of certain molecules may follow a very specific path, but the molecules are not capable of breaking apart to produce two identical copies of themselves.

   In biological evolution, the cells are able to divide and produce two identical copies of themselves. The process is not at all random and is controlled by information stored in the cells DNA.

**WORKED EXAMPLES OF PROBLEMS REQUIRING MATHEMATICS**

1. If a human generation, defined as the time from birth to childbearing, is 20 years, how many generations have there been since the beginning of human history 3 million years ago?

   Solution:
   We know
   the time for one generation = 20 years and
   the age of humankind = 3 million years = 3,000,000 years.

   We can find the number of generations by dividing the amount of time that humans have been around by the time for one generation.

   $$number\ of\ generations = \frac{age\ of\ humankind}{time\ for\ one\ generation} = \frac{3,000,000\ years}{20\ years} = 150,000\ generations$$

   There have been approximately 150,000 generations of humans in the last 3 million years.

2.  If stars in the vicinity of the sun are on average 5 light years apart, approximately how many stars are their within 200 ly of Earth?

Solution:
We know
the average distance between stars = $r$ = 5 light years = 5 ly and
we want to know how many are within a distance = $R$ = 200 ly of us.

The number of stars within 200 ly of the sun can be found by dividing the volume of space within 200 ly of the sun, $V_T$, by the volume of space around an individual star, $V_1$.

We can think of each star as being surrounded by its own little bubble of space. The volume of each star's bubble, $V_1$, is equal to the volume of a sphere with a radius of 2.5 ly, that is, one-half of 5 ly.

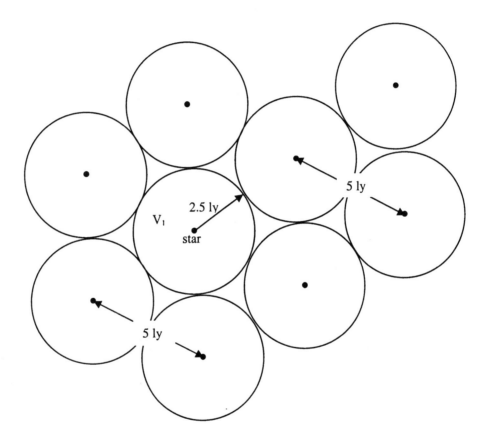

The volume of space around each star is given by $V_1 = \frac{4}{3} \pi r^3 = \frac{4}{3} \pi (5 \text{ ly})^3$.

The total volume of space within 200 ly is given by $V_T = \frac{4}{3} \pi r^3 = \frac{4}{3} \pi (200 \text{ ly})^3$.

As we said earlier, the number of stars within a given volume is equal to that volume divided by the average volume surrounding a star.

$$number\ of\ stars = \frac{V_T}{V_1} = \frac{\frac{4}{3} \cdot \pi \cdot R^3}{\frac{4}{3} \cdot \pi \cdot r^3} = \frac{R^3}{r^3} = \frac{200^3}{5^3} = 64{,}000 \text{ stars}$$

There are approximately 64,000 stars within 200 light years of the sun.

3. What is the mass of a star whose main sequence lifetime is 3.5 billion years?

Solution:

We know the main sequence age of the star = 3.5 billion years = $3.5 \times 10^9$ years.

We can find the mass of the star using *By the Numbers 8-5* on page 151.

$$T_{years} = \frac{1 \times 10^{10}}{M^{2.5}}.$$

We want to solve this for $M$, so multiply both sides by $M^{2.5}$

$$M^{2.5} \cdot T_{years} = M^{2.5} \cdot \frac{1 \times 10^{10}}{M^{2.5}} = 1 \times 10^{10}$$

Now divide both sides by $T_{years}$.

$$\frac{M^{2.5} \cdot T_{years}}{T_{years}} = \frac{1 \times 10^{10}}{T_{years}}$$

The $T_{years}$ cancels on the left side so that

$$M^{2.5} = \frac{1 \times 10^{10}}{T_{years}} = \frac{1 \times 10^{10}}{3.5 \times 10^9} = 2.86$$

$$M = \sqrt[2.5]{M} = \sqrt[2.5]{2.86} = 1.5 \text{ solar masses}$$

The last step requires that you take the 2.5-root of 2.86. This is done with the key labeled $\boxed{x^y}$ or $\boxed{x^{1/y}}$ both of which can be used to find the answer.

To use the $\boxed{x^{1/y}}$ key you may have to first push a $\boxed{\text{second}}$ or a $\boxed{\text{function}}$ key. Begin by entering 2.89, then activate the $\boxed{x^{1/y}}$ key, again this may require first pushing push a $\boxed{\text{second}}$ or a $\boxed{\text{function}}$ key. Now enter $\boxed{2.5}$, and press the $\boxed{=}$ key. The display will read $\boxed{1.52246183}$. This is rounded to 1.5

To use the $\boxed{x^y}$ key, you need to realize that $\frac{1}{2.5} = 0.4$. First enter the number 2.86, then push $\boxed{x^y}$, enter 0.4, and press the $\boxed{=}$ key. The display will read $\boxed{1.52246183}$. This is rounded to 1.5.

A star with a main sequence lifetime of 3.5 billion years, would have a mass of 1.5 solar masses.

## MORE MATH RELATED PROBLEMS
(Answers at the end of the chapter)

1. What is the lifetime of a 3 solar mass main sequence star?

2. What is the mass of a star whose main sequence lifetime is 2 billion years?

3. If the average human generation is 20 years, how many generations have there been since Galileo first observed the moons of Jupiter around 1605?


## PRACTICE TEST
### Multiple Choice Questions

1. What is the significance of the Miller experiment?
   a. It showed that simple life forms could change from one species to another.
   b. It showed that DNA can replicate itself.
   c. It showed that basic chemicals stimulated with electrical sparks could produce amino acids.
   d. It showed that the DNA molecule is extremely fragile.
   e. It showed that life must be based on carbon molecules.

2. The portion of the electromagnetic spectrum between the 21 cm line of neutral hydrogen and the 18 cm line of OH is referred to as the
   a. life zone.
   b. water hole.
   c. primordial soup.
   d. Cambrian zone.
   e. alien window.

3. The most important quantity necessary for life to form appears to be the existence of
   a. silicone in the planet's atmosphere.
   b. meteorites that carry amino acids to the planet's surface.
   c. carbon dioxide in the planet's atmosphere.
   d. liquid water on the planet's surface.
   e. strong electrical storms in the planets atmosphere.

4. Complex life first appeared on Earth during the Cambrian Period, about
   a. 0.5 billion years ago.
   b. 2 million years ago.
   c. 100 thousand years ago.
   d. 4 billion years ago.
   e. 10 million years ago.

5. DNA in a nucleus contains
   a. the information needed to select mutations that are beneficial to the organism.
   b. the information needed to guide all processes in an organism.
   c. nutrients to feed the cell.
   d. all of the above.
   e. none of the above.

6. The building blocks of life are relatively simple chemicals known as
   a. the primordial soup.
   b. RNA.
   c. chromosomes.
   d. Cytosines.
   e. amino acids.

7. Spectral type G and K stars are expected to be the best places to find life around other stars because
    I.   They have high enough temperatures to maintain a significant life zone.
    II.  They produce large amounts of ultraviolet radiation.
    III. They have main sequence lifetimes over 4 billion years.

    a.  I & II
    b.  I & III
    c.  II & III
    d.  I, II, & III
    e.  none of them

8. Scientific explanations are based on fundamental rules of
    a.  evidence and intellectual honesty.
    b.  scientific faith.
    c.  skepticism.
    d.  mathematics.
    e.  the Geneva convention.

9. The Viking spacecraft searched for life
    a.  near volcanic vents on the ocean floor of Earth.
    b.  under the frozen oceans on Europa.
    c.  in the atmosphere of Jupiter.
    d.  in the soils of Mars.
    e.  in the ice on the moon.

10. Mars, Europa, and Titan have all been considered as possible locations in the solar system where rudimentary life forms like bacteria may have formed or may be forming. Why are these three objects worth a closer look?
    a.  They are all inside the life zone of the sun.
    b.  They all show evidence of unnatural changes on their surface.
    c.  They all have a magnetic field about as strong as Earth.
    d.  They all posses ozone in their atmosphere.
    e.  They all show evidence for the existence of liquid near the surface currently or in their recent past.

**Fill in the Blank Questions**

1. _____ is the molecule within a cell that actually assembles proteins from amino acids.

2. DNA molecules contain patterns for the production of _____.

3. All life on Earth is based on _____ chemistry.

4. The _____ equation provides an estimate of the number of technological civilizations with which we might be able to communicate.

5. The region around a star where the temperature permits the existence of liquid water is known as the _____ _____.

**True-False Questions**

16. The stars most likely to have inhabited planets are G to K main sequence stars.

17. At present bacterial life on Earth is believed to have begun on the continent of Africa.

18. An offspring born with altered DNA due to radioactivity, cosmic rays, or errors in reproduction is much more likely to survive than if its DNA were not altered.

19. The Miller experiment succeeded in creating amino acids.

20. Massive main sequence stars are not likely to have planets that contain life because these stars are too hot to allow water to exist as a liquid on any planets that might form.

## ADDITIONAL READING

Crawford, Ian "Where are They?" *Scientific American 283* (July 2000), p. 38.

Grinspoon, David "SETI and the Science Wars." *Astronomy 28* (May 2000), p. 52.

Rees, Martin "Is Earth Unique?" *Astronomy 29* (March 2001), p. 54.

## ANSWERS TO MORE MATH RELATED PROBLEMS
1. $6.4 \times 10^8$ years = 640 million years
2. 1.9 solar masses
3. 20 generations

## ANSWERS TO PRACTICE TEST

| | | |
|---|---|---|
| 1. c | 8. a | 15. life zone |
| 2. b | 9. d | 16. T |
| 3. d | 10. e | 17. F |
| 4. a | 11. RNA | 18. F |
| 5. b | 12. protein | 19. T |
| 6. e | 13. carbon | 20. F |
| 7. b | 14. Drake | |

# APPENDIX 1

# SCIENTIFIC NOTATION

## INTRODUCTION

This appendix is intended to (re)introduce you to the concepts of scientific notation and how to use your calculator to do perform multiplication and division using scientific notation. These discussions are intended to help the student who has not used math for some time or who is apprehensive about math.

Scientific notation is nothing more than a method for writing down extremely large or small numbers. Extremely large numbers like 3,150,000,00, and extremely small numbers like 0.00000000783 can be a pain to deal with. It is easy to forget to write down one of the zeros and it is tedious simply to record them all. In astronomy we deal with numbers that are both extremely large and extremely small. Therefore, scientific notation can help us accurately keep track of the numbers, and it provides a simplified mechanism for us to compare the relative sizes of numbers.

The basis of scientific notation is that 10 multiplied times its self a number of times is always 1 followed by a string of zeros, (e.g. 10×10×10 = 1000, 10×10×10×10×10×10 = 1,000,000). It is easy to see that if you count the number of 10s that are multiplied together your answer is 1 followed by that number of zeros (e.g. 10×10×10×10 = 10 times itself 4 times = 1 followed by 4 zeros = 10,000). When one wants to multiply a number times itself several times we usually use exponents, that is if we want to multiply 10 times itself 4 time we don't usually write 10×10×10×10, instead we would write $10^4$ (read ten to the fourth power, or simply ten to the fourth). With this we can see that

$$10 = 10^1$$

$$100 = 10 \times 10 = 10^2$$

$$1000 = 10 \times 10 \times 10 = 10^3 \text{ and}$$

$$1,000,000,000,000 = 10 \times 10 \times 10 \times 10 \times 10 \times 10 \times 10 \times 10 \times 10 \times 10 \times 10 \times 10 = 10^{12}.$$

Similarly, when the number is smaller than 1 we use negative numbers with the exponent:

$$1 \div 10 = \frac{1}{10} = 0.1 = 10^{-1}$$

$$1 \div 100 = \frac{1}{100} = 0.01 = 10^{-2}$$

$$1 \div 1,000 = \frac{1}{1,000} = 0.001 = 10^{-3}$$

$$1 \div 1,000,000,000,000 = \frac{1}{1,000,000,000,000} = 0.000000000001 = 10^{-12}$$

Notice that with numbers smaller than 1, what you need to do is count how many places you have to move the decimal point to get it just to the right of the one. In the case of $10^{-12}$ above, the decimal gets moved past 11 zeros and the 1 so it is moved 12 places.

Note that $10^1 = 10$ and $10^{-1} = 0.1$, and we define $10^0 = 1$.

Well this is fine for multiples of 10 but what about numbers like 3,150,000,000? We could write this number as $3.15 \times 1,000,000,000$, but $1,000,000,000 = 10^9$. Therefore, we can rewrite 3,150,000,000 as $3.15 \times 10^9$, where now I understand that this means to move the decimal 9 places to the right of where it is i.e. $3.15 \times 10^9 = 3,150,000,000$.

Some examples:

$$54,678 = 5.4678 \times 10^4$$

$$892,345,518 = 8.92 \times 10^8$$

$$0.000325 = 3.25 \times 10^{-4}$$

$$0.00000000000008378 = 8.378 \times 10^{-14}$$

In essence, scientific notation boils down to counting the number of places you need to move the decimal point. However, scientific notation also makes arithmetic a lot easier when dealing with very big or very small numbers. With electronic calculators this isn't as big an advantage as it used to be. However, some numbers cannot be entered into a calculator without using scientific notation, and many numbers are simply faster to enter in scientific notation.

## SCIENTIFIC NOTATION AND YOUR CALCULATOR

Many students make mistakes entering scientific notation numbers into their calculators. Exactly how you should enter numbers in your calculator depends on the make and model of the calculator that you are using. The keys you need to push and the display that results vary significantly from calculator to calculator. If you are having trouble with your calculator, I suggest that you consult the instructions that came with it, or ask a friend who seems to understand her/his calculator to give you a hand. In the discussion below, I am going to give you some examples of keys you need to push and the displays that will result. I will use a thick box around all $\boxed{\textbf{displays}}$, and a thin box around all $\boxed{\text{key labels}}$.

Scientific numbers can be displayed in several different formats. On a calculator the number $3.6 \times 10^{15}$ might be displayed as $\boxed{\textbf{3.6  15}}$, $\boxed{\textbf{3.6 E15}}$, $\boxed{\textbf{3.6 } \times 10 + 15}$, or $\boxed{\textbf{3.6}^{15}}$. To find out the display that your calculator uses, you need to enter a number in scientific notation. Most calculators have a key for entering scientific notation labeled with one of three different labels, either $\boxed{\text{EE}}$, $\boxed{\text{EXP}}$, or $\boxed{\text{EEX}}$. In some cases you will need to press a $\boxed{\text{function}}$ key or a $\boxed{\text{2nd}}$ key to activate this key. If you want to enter the number $3.6 \times 10^{15}$, you use the following key sequence

$$\boxed{3} \quad \boxed{.} \quad \boxed{6} \quad \boxed{\text{EE}} \quad \boxed{1} \quad \boxed{5}$$

Your display should read $\boxed{\textbf{3.6  15}}$ or one of the other displays above.

Let's try a multiplication and a division problem to make sure everything is working correctly. We want to multiply $4.89 \times 10^{22}$ by $8.41 \times 10^9$. The key sequence on most calculators would be:

| 4 | . | 8 | 9 | EE | 2 | 2 | × |

| 8 | . | 4 | 1 | EE | 9 | = |

The display should be | 4.1125  32 |

So $4.89 \times 10^{22}$ times $8.41 \times 10^9$ equals $4.11 \times 10^{32}$.

Division is exactly like multiplication, except we have to push the | ÷ | key. As an example find the answer to

$$\frac{6.52 \times 10^{14}}{7.39 \times 10^5}.$$

The following key sequence should provide the correct answer.

| 6 | . | 5 | 2 | EE | 14 | ÷ |

| 7 | . | 3 | 9 | EE | 5 |

The display should read | 8.8227  08 |.

So $\dfrac{6.52 \times 10^{14}}{7.39 \times 10^5} = 8.82 \times 10^8$.

If you aren't getting the answers given above, then you need to talk to a tutor, lab assistant, math-whiz friend, or your professor to get things working correctly.

There is one more little wrinkle that needs to be addressed, and that is entering very tiny numbers like $0.00000000000274 = 2.74 \times 10^{-12}$. The problem often occurs in getting the minus sign entered properly. Most calculators have a key labeled | ± | or | +/_ | that you will use. The key sequence needed to enter $2.74 \times 10^{-12}$ is as follows.

| 2 | . | 7 | 4 | EE | ± | 1 | 2 |

The display should read | 2.74  −12 |.

Try this one. $6.42 \times 10^{-6}$ times $5.39 \times 10^8$. The key sequence is

| 6 | . | 4 | 2 | EE | ± | 6 | × |

| 5 | . | 3 | 9 | EE | 8 |.

The display should read | 3.4604  03 | or | 3460 |.

We find that $(6.42 \times 10^{-6}) \cdot (5.39 \times 10^8) = 3.46 \times 10^3$, which is equal to 3,460.

Try an example involving division. Calculate

$$\frac{4.38 \times 10^8}{2.77 \times 10^{-17}}.$$

The key sequence is

| 4 | . | 3 | 8 | EE | 8 | ÷ |

| 2 | . | 7 | 7 | EE | ± | 1 | 7 | = |.

The display should read $\boxed{1.5812 \quad 25}$.

We find that $\dfrac{4.38 \times 10^8}{2.77 \times 10^{-17}} = 1.58 \times 10^{25}$.

Here are some problems to practice on, with answers at the bottom of the list.

a)   $4.2 \times 10^5 \cdot 3.0 \times 10^2 = $ _____

b)   $5.25 \times 10^{-6} \cdot 8.3 \times 10^9 = $ _____

c)   $6.11 \times 10^{21} \cdot 7.42 \times 10^{31} = $ _____

d)   $1.8 \times 10^{-17} \cdot 2.4 \times 10^{-5} = $ _____

e)   $9.0 \times 10^6 \div 3.0 \times 10^3 = $ _____

f)   $7.18 \times 10^{-4} \div 6.3 \times 10^8 = $ _____

g)   $2.36 \times 10^4 \div 4.51 \times 10^{-12} = $ _____

h)   $5.22 \times 10^{-15} \div 2.61 \times 10^{-9} = $ _____

**Answers**

a)  $1.26 \times 10^8$    b)  $4.34 \times 10^4$    c)  $4.53 \times 10^{53}$    d)  $4.32 \times 10^{-22}$    e)  $3.0 \times 10^3$
f)  $1.14 \times 10^{-12}$    g)  $5.23 \times 10^{15}$    h)  $2.0 \times 10^{-6}$

# APPENDIX 2

# ASTROLOGY, UFOs AND PSEUDOSCIENCE

## by Michael Seeds

By taking a single course in astronomy, you have become, compared with the average person, an astronomer. With this knowledge comes the responsibility that all astronomers feel—the responsibility to help others understand the science of astronomy and evaluate such false sciences as astrology and UFOs.

In this essay we will examine such false sciences. We will try to understand why they are not scientific and why people believe in them anyway. Most of all, we will try to understand how we as astronomers can help people distinguish good science from bad.

### ASTROLOGY

Astrology is an ancient superstition that holds that a person's personality and life are influenced by the position of the sun, moon, planets, and stars at the moment of birth. In addition, astrologers claim that the daily changes in the location of the heavenly bodies can influence events in our lives. All of this is summarized in a horoscope, a diagram of the zodiac, showing the position of the heavenly bodies.

This belief originated over 3000 years ago in the sky-worshipping religions of Babylonia and was later modified by the Egyptians and Greeks. All these cultures believed that their gods were manifested in the moving heavenly bodies, so it was reasonable for them to imagine these bodies could affect lives on Earth. Modern people no longer worship the ancient sky gods, but many people today still profess a belief in the powers of those gods as represented in astrology. *

Although the ancient religions have been abandoned for thousands of years, astrology has survived nearly unchanged since the second century AD. Perhaps it has survived because it appeals to our deepest fears and needs.

### Does Astrology Work?

We must consider two questions. First, is there any physical reason we should expect astrology to work? Second, is there any evidence that it actually does work?

According to the basic claim of astrology, a baby's character, personality, and future life are affected in some way by the location of the sun, moon, and planets along the zodiac at the moment of birth. In fact, there seems to be no way for these objects to act on the body of the baby. The force of gravity is the most likely agent, but the gravitational field of the doctor is much larger than that of the planets. Thus it would seem more important where the doctor stands than where Mars is along the zodiac.

If gravity is not the agent of influence, we have only a few other effects to consider. Light and other electromagnetic radiation from the sun, moon, and planets do not generally enter hospital delivery rooms, and the strong nuclear force, and the weak nuclear force are even less effective than gravity. No other forces, rays, or influences in nature are known, so there is no physical agency to explain the claims of astrology, and we have no physical process to explain astrology.

Most astrologers ignore this problem, but a few propose influences that go beyond physical processes. The moon, for example, causes the tides, so ancient astrologers proposed that the moon rules all fluids including those of our bodies. This reflects a Middle Age view of nature as governed by hidden purpose and undetectable sympathies and antipathies. In fact, we now know that the tides are caused by gravity, and that the tides in an object as small as a human are astonishingly small, far too small to have any effect. Modern science has long abandoned the mystical naturalism of the Middle Ages, and that leaves no physical force to carry the influences attributed to astrology. There is no reason we should expect astrology to work.

Our second question is more direct. Is there any evidence that astrology does work? In fact, astrology has been tested many times over the centuries, and no legitimate test has proven any astrological influence.

A few positive tests have been reported by the grocery store tabloids, but when these tests were repeated by trained statisticians, astrology failed. Extensive tests have searched in vain for relationships between birth sign and such personal characteristics as blood type, bust size, handedness, profession, extroversion/introversion, divorce rate, and others. Astrology would be very useful if it worked, but no relationship has ever been found.

## Should We Respect Astrology?

As an astronomer, you must decide how to view astrology. Should you respect it as an alternative scientific view, as a religion, or as a harmless superstition?

Some astrologers claim that astrology is a science. They claim that it is based on scientific methods and quote elaborate theories about unknown rays from the planets. If this were true, astrology would deserve our respect as a science, but in fact there is no scientific evidence that astrology is valid. In fact, all scientific evidence argues against astrology.

In addition, astrology is not based on scientific methods. A scientist, above all else, must be willing to revise hypotheses in the face of contradictory evidence. Scientific theories must always be open to revision, but the principles of astrology have not been revised for almost 2000 years. For instance, precession has altered the location of the vernal equinox among the constellations so that the astrological signs of the zodiac are now one entire constellation out of place. For another example, three new planets have been discovered since astrology was founded. Astrology has not been revised to adapt to such changes or to contradictory evidence from tests of astrology. A theory that is not open to revision, cannot be scientific. Even if we interpret *scientific* to mean *systematic,* astrology is not scientific. Most people meet astrology in newspapers and magazines, and those predictions cannot have been "cast" according to systematic rules. To cast your horoscope, an astrologer would have to know the date, time, latitude, and longitude of your birth. Even if a newspaper horoscope did happen to be correct for you, it could be correct for no other person in the world.

We have only to examine the astrology in newspaper columns to see that it cannot have been derived from a systematic set of rules. The following predictions for Libra were extracted from newspapers on January 11, 1984. Both the zodiacal sign and the date were selected at random.

*Gain indicated through written word. Locate legal documents, special manuscripts and be aware of rights, permissions. –Sydney Omarr*

*As best you can sort out the pieces. – Clay R. Pollan*

*Exuberant moods sweep you up in a whirl of excitement. Make sure all financial negotiations are on the up and up. The evening promises much love and affection. –Joyce Jillson*

*Renew worthwhile agreements with partners. A worldly matter could be confusing early. Be dynamic. –Carroll Righter*

*You're able to lift the spirits of a loved one. Your belief in your own potential rubs off on others. Enter financial negotiations. –Frances Drake*

*Stick close to home today. Those who can use public transportation should do so. Encourage teenagers to earn their own pocket money. –Jeane Dixon*

Like all astrological predictions, these are vague, generally optimistic, give us good advice, and refer to common features in our daily lives such as friends, loved ones, financial matters, jobs, and so on. Even so, the disagreement among these predictions shows that astrology cannot claim to be systematic. Even the most popular astrologers cannot agree whether we should be exuberant and dynamic or stay home and study legal documents.

Astrology cannot claim to be a science. It is not open to revision as dictated by experiment, nor is it systematic and consistent in its methods. If someone tells us astrology is a science, we have every right to withhold our respect.

But perhaps we should respect astrology as a religion. Certainly the most devout believers in astrology have a faith in its principles that approaches religious intensity, and astrology did originate in ancient sky worship. Yet astrology is a false religion because it does not incorporate any standards of moral behavior. Indeed, true believers in astrology might steal and murder and then blame their crimes on the chance alignments of the stars at their birth. A true religious faith in astrology dilutes the responsibility we feel for our own actions.

If we astronomers cannot respect astrology as a science or as a religion, then perhaps we should respect it as an ancient and harmless superstition. After all, we knock on wood for luck, and astrology is only another from of appealing to gods no longer worshiped. Yet astrology is not a harmless superstition. It asks us to believe that our personalities are fixed at the moment of birth. If that is true, then our fate is sealed and we who are shy can never make many friends. We who are weak can never grow strong. We who are sad can never be happy.

Worse yet, astrology as a superstition tells us that the daily events of our lives are beyond our control. If our love life falters one day, it is hopeless to try to save it. If our studies go well, it is not to our credit but is merely our stars at work. We are held hostage by our stars. In its most intense form, a superstitious belief in astrology is depersonalizing, and those who tell us they believe in astrology as a superstition deserve our sympathy rather than our respect.

### An Attitude Toward Astrology

Although each astronomer must develop an independent personal attitude toward astrology, it does seem clear that we cannot respect it as a science, as a religion, or as a superstition. Perhaps we can view it as a deep and ancient expression of our human need to believe that our lives have meaning beyond our daily existence. However inevitable astrology is, we should not be content with its survival. It is an ancient belief that does not work. It is not science, not religion, but a potentially harmful superstition. Our universe harbors greater elegance and greater power than those forces astrology purports to describe.

## UFO'S AND VISITORS FROM THE STARS

If your friends discover that you know astronomy, they may ask you about unidentified flying objects (UFOs) and visitors from outer space. Perhaps because astronomers watch the sky, your friends may expect you to be an expert on this subject, so we must try to decide what to say about UFOs.

### The UFO Principle

Those who believe in UFOs believe in something we might call the UFO principle—UFOs are spacecraft piloted by creatures from another world. We must ask ourselves if this is a reasonable belief and what evidence exists to support it.

As astronomers, we know that life could be very common in the universe. Planets appear to be by-products of star foundation, and it does seem possible that life could originate on any world where conditions were right. Once active, life modifies itself to fit its environment, so although we don't know how common suitable planets are, we must admit that our galaxy could be well populated. Could such worlds, however, send spacecraft to visit ours?

No other planet in our solar system is suitable for intelligent life, so any visitors to Earth must come from other planetary systems orbiting other stars. This argues against the UFO principle because of the difficulty in traveling between stars. Other creatures might be different from us, but they must still obey the same laws of physics, and that means that traveling from their star system to ours would be almost impossible. The stars are very, very far apart.

Thus the astronomy we know tells us that the UFO principle is unreasonable. Not because we don't believe in aliens on other worlds, but because we suspect they could never visit us. But we cannot discard a hypothesis because we find it unreasonable. We must look at the evidence of UFO sightings.

### Sorting Sightings

We can divide UFO sightings into three categories—hoaxes, natural phenomena, and the rest. To understand these sightings, we must think about the people who report them.

A few UFO sightings are hoaxes. These are usually the most dramatic, get the most press coverage, and usually lead to books, lectures, and magazine articles. We must not forget that many people earn a good living from UFOs and that a successful hoax is worth a lot of money to many people.

Most UFO sightings are made by honest people who mistake natural phenomena for something peculiar. The brightest stars and planets, when seen near the horizon, twinkle so violently they can appear to be flashing colored lights and rotating. Observatories receive many such reports when Venus is brilliant in the evening or morning sky. Other objects mistaken for UFOs include airplanes, weather balloons, clouds, birds, and so on.

Once we discount hoaxes and mistakes, we have only a few sightings left. These are the ones that cannot be explained so easily but do appear to be honest reports. Some of these are misinterpretations of original descriptions. For example, Uncle George tells us he saw a bright light above his barn (Venus?), and we tell a reporter that our uncle saw a bright light hovering above his barn, and the reporter prints a story about a brilliant object hovering above a barn. We all tend to exaggerate a story to make it just a bit better than the actual event. In the telling of the story, fishes get bigger and UFOs get more elaborate.

Some of the remaining sightings can be attributed to a well-known psychological principle—we see what we expect to see. Thus people who read about UFOs regularly are more likely to see one. Also, if we view an event and in the first split second misidentify a meteor as a spaceship, our brain quickly fits additional details into the spaceship pattern, and we think we see a glowing spaceship with illuminated ports and a flaming exhaust. It is quite possible for an honest person to describe a natural event in such a way that it cannot be identified with any natural phenomenon.

Thus we should expect any large set of observations to contain some residual of unexplainable reports. But that does not prove that none of those reports is valid. To test such a conclusion, we would have to study all reports in great detail, and a number of teams of scientists have done exactly that. None of these studies has found even one sighting that is conclusive. A few defenders of the UFO principle claim that the government, the military, and the scientific community are covering up the truth in a grand conspiracy. This hypothesis is very difficult to accept. Recent history tells us that governments are not very good at covering things up, even when only a few people know the truth, and a grand conspiracy to hide the truth about UFOs would involve the government, the military, the press, the airline industry, and the scientific community not only in the United States but all around the world. A visit from aliens from another world would be the news story of all time, the scientific discovery to assure a scientist of everlasting fame. It is very difficult to believe that such a conspiracy of silence could survive.

In fact, a visit from people beyond Earth would be a wonderful experience for our planet. Such aliens might exist, and it is possible though highly unlikely that they could travel to our solar system. But at present there is not a single observation to support the idea that UFOs are spacecraft from other worlds.

Still some people believe in UFOs in spite of the evidence, and they believe for very understandable reasons. We all like to believe that greater powers are watching over us, and psychologists know that UFO sightings always increase in times of public stress. Sightings in Russia, for example, shot up with the fall of communism. Also, everyone likes to be among that special group that is in the know. We feel set apart when we think we know the truth being hidden from the common people by some grand conspiracy. Of course, most people enjoy a scary story, and UFOs and space aliens are fun to think about. The psychological need to believe in UFOs is complex and deep seated, but as astronomers we must accept the rule of evidence. However exciting UFOs might be, there is no rational reason to believe in the UFO principle.

## SCIENCE AND PSEUDOSCIENCE

UFOs and astrology are only two of the subjects that we can describe as pseudosciences. A *pseudoscience* is something that pretends to be a science but does not obey the rules of good conduct common to all sciences. Thus such subjects are false sciences.

True science is a method of studying nature. It is a set of rules that prevents scientists from lying to one another or to themselves even by accident. Hypotheses must be open to testing and must be revised in the face of contradictory evidence. All evidence must be considered, and all alternative hypotheses must be explored. The rules of good science are nothing more than the rules of good thinking—that is, the rules of intellectual honesty.

You have probably read of some subjects that are pseudosciences. A few years ago, popular paperbacks claimed that all the planets would line up and Earth would be shocked by earthquakes and storms. Nothing happened. A few years before that, popular books claimed that ancient astronauts (aliens from another world) had visited Earth and taught primitive humans how to be civilized. A few decades ago, a popular book argued that many of the events in the Bible could be explained if Venus had been expelled from Jupiter and had collided with Earth. You can still buy paperbacks claiming that Earth is hollow and inhabited on the inside by a super race. Of course, every generation has its pseudoscientists who claim to cure our worst diseases with quack treatments. Quite a number of people make their living promoting pseudoscientific ideas.

When you meet such a fantastic hypothesis, ask yourself if it is pseudoscience. Does the author discuss all of the relevant evidence or only that which supports the hypothesis? Does the author consider all alternative hypotheses? That is, is the author being intellectually honest? Such questions can usually separate pseudoscience from real science.

Ultimately, you and I must reach our own conclusions, but we can deal most effectively with astrology, UFOs, and other pseudosciences if we remember that people often believe in things for reasons that go beyond the rules of good logic. If we are to be astronomers and help others understand their place in the universe, we must help them use the same rules of intellectual honesty that we demand of ourselves.

*As recently as January 1984, the Vatican daily newspaper urged Roman Catholics not to believe in astrology lest they "subject the faith to a risk of defilement."

Note: This article was originally published in *Astronomy Insights: Study Tips, Plus a Look at Astrology, UFOs*, by Michael Seeds, Belmont, CA: Brooks Cole, 1996.